FAO Agricultural Policy and Economic Development Series
4

Guidelines for the integration of sustainable agriculture and rural development into agricultural policies

by
J. Brian Hardaker

FOOD AND AGRICULTURE ORGANIZATION OF THE UNITED NATIONS
Rome, 1997

CONTENTS

SUMMARY

The concept of SARD

The FAO Council has defined sustainable agriculture and rural development (SARD) as:

> ... the management and conservation of the natural resource base, and the orientation of technological and institutional change in such a manner as to ensure the attainment and continued satisfaction of human needs for present and future generations. Such sustainable development (in the agriculture, forestry and fisheries sectors) conserves land, water, plant and animal genetic resources, is environmentally non-degrading, technically appropriate, economically viable and socially acceptable (FAO 1989).

The focus is on the sustainable welfare of humans, living now and those not yet born. In other words, it is about sustainable livelihoods, now and in the future.

A principle of sustainability is that we pass on to the next generation a stock of resources that is at least as productive as the stock we have today. However, since it is neither necessary nor rational to strive to make every sub-system sustainable, this principle needs to be applied at the highest relevant system level. It is also not necessary and not even possible to pass on an enhanced stock of every type of resource. Substitution of resources will be possible as future generations substitute resources that are relatively more abundant for those that have grown scarce. In agriculture, the substitution of human capital—in the shape of knowledge about improved technologies—for land and labour, has been important in the past and will be at least as important in the future.

Policy objectives

Policy objectives for SARD may be summed up as the pursuit of the goals of growth, equity, efficiency and sustainability. Growth is important to meet the food needs of growing populations with rising incomes and to provide continued sustainable livelihoods for rural people in the future. Equity is important in terms of the relief of poverty and deprivation for this and future generations. Efficiency matters since we cannot afford to waste resources. Finally, sustainability is the objective that has come into increased prominence with the recognition of the significant threats that exist to future welfare and the environment. Sustainability has many dimensions and interpretations but, in the context of agriculture, embraces food security, responsibility in resource use and environmental management, and the resilience of production systems to shocks and challenges.

There is interdependence between each of these four objectives, so that the pursuit of SARD requires an integrated approach to policy making in which all four aspects are considered.

Policy framework

There needs to be an appropriate framework in place for the formulation and effective implementation of policies for SARD. Elements of this framework include:

- Taking a long-term, global perspective.
- Implementing complementary sectoral and macro-economic policies.
- Developing coordinated and consistent policies within the agricultural sector.
- Having an appropriate legal and institutional structure in place.
- Following an approach in which the inherent uncertainties of planning for SARD are recognized and accommodated.
- Taking an adaptive, inter-disciplinary and multi-disciplinary 'systems approach' to policy planning.

Why policy intervention is necessary

Policy intervention for SARD is necessary because of market and policy failures. Markets fail when prices do not reflect the real values of resources, goods or services. That means that producers and consumers get the wrong signals about relative scarcities. Sources of market failure include attenuated property rights, externalities, imperfect information, monopolistic competition, and imperfect and distorted capital markets. A form of market failure especially important for SARD is unequal market power, whereby the needs of the poor are swamped by the greater purchasing power of the rich. This inequality is especially severe in relation to the needs of poor people yet unborn and substantially neglected in today's markets.

The existence of market failure does not automatically justify policy interventions. Policy interventions are often ill-conceived or inadequately implemented so that their impacts are at best ineffective and at worst may be contrary to intentions. Command and control measures directed towards SARD have a poor record of success. When intervention is needed, therefore, it is best to look first at changing incentives to encourage behaviour in line with the sustainability principle. On occasion, of course, control measures, or even direct action by government agencies, may be the best approach, perhaps the only one possible.

Strategic issues

A strategy is the way the policy campaign is managed. The following steps should be kept in mind when outlining such a strategy:

- Consistency with national objectives, including the identification and, where possible, the resolution of conflicting objectives.

- Identifying overall goals of strategies, consistent with national goals, but narrowed to be operationally meaningful.

- Identifying components of the strategy, meaning the bundle of measures that must be taken to attain the stated objectives.

- Identifying areas of intervention, that is, deciding what is the proper role for government in promoting SARD.

- Identifying investment strategies, which follows from the previous point; many forms of intervention will require capital injection into, say, infrastructure, and it will be necessary to plan the nature, scale and timing of the required investments.

Key strategic issues to be addressed include: the emphasis to be placed on economic growth versus other objectives that may be seen as more directly related to SARD; the balance between reliance on market-based incentives versus command and control measures; and the priority to be given to 'top down' versus 'bottom up' approaches.

While growth of agriculture is necessary for SARD, the choice of a growth strategy requires policy attention to the question of how that growth can be achieved without degrading agricultural resources or the environment. Specific practices and approaches, such as integrated pest management and responsible fisheries may allow more intensive and more sustainable production. Such approaches need to be identified, developed and promoted. Attention also needs to be given to opportunities to exploit substitution possibilities in production in order to account for relative resource scarcities.

A key element of a strategy for SARD is to try to provide the appropriate incentives by getting prices right. Sometimes it is possible to devise policy instruments to substitute for missing or imperfect markets, to create markets where none existed before, or to make existing markets work better. However, where such interventions are not possible, it is still important that policy makers seek to use the 'right' prices in evaluating the impacts of interventions. Essentially, that means using prices in agricultural project appraisals that reflect, so far as possible, the true costs and benefits of the alternatives considered to present and future generations. Similarly, resource accounting methods have been developed to give a better representation in national income calculations of changes in levels of resource stocks and environmental quality.

SARD is not wholly or even mainly a matter for policy makers. In most countries, it is the farmers, forest people and fisherfolk who are responsible for most of the resource utilization decisions that affect SARD. Their participation is therefore vital. A strategy for SARD must include measures to promote this participation and to create conditions for people to change to more sustainable ways. Moreover, by encouraging the participation of rural people, policy makers in government can learn from them about their needs and circumstances, and about their likely responses to policy initiatives. Policy makers need to find culturally acceptable and effective ways of involving people in their decisions and of motivating people to search for and adopt their own solutions to problems of unsustainability.

Policy instruments

There are five main types of policy instruments that affect SARD:

- general economic and social policies;

- policies relating to agricultural and rural development;

- policies relating to markets, including the establishment of market institutions and rules, and to allocation of property rights;

- policies aimed at establishing democratic and participatory processes; and

- policies designed specifically to influence natural resource use and protect the environment.

General economic and social policies. Under this heading, fiscal and monetary policies can influence SARD in many ways. Many governments, especially in industrialized countries, raise the majority of their taxes from progressive income taxes that, indirectly, increase the cost of labour to employers. As a result, firms will find it profitable to substitute other, less heavily taxed natural resources for labour, leading to a too high utilization of natural resources relative to labour. Biasing the taxation system more strongly towards natural resource taxes may help promote more sustainable development.

Prudent fiscal management is important for SARD since economic instability is likely to discourage private investment, and frequent budget crises may disrupt the provision of important public services. A too large public sector may 'crowd out' development in private sector activities, including agriculture.

Low interest rates will favour long-term investments in resource conservation measures. However, low interest rates achieved by artificially holding down rates may inhibit development of the financial sector, discouraging formal savings and limiting the funds available for lending for rural (and other) development.

Trade and exchange rate polices that distort prices of agricultural inputs and products can have negative impacts on SARD. For instance, some countries have sought to maintain a high official exchange rate by limiting imports. Such policies turn the domestic terms of trade against sectors producing tradeable goods, including agriculture. The distortions induced can be massive, and devastating for SARD.

The sustainability principle of transferring the equivalent of the current resource base to the next generation means that investment policy is crucial. Policy instruments for SARD include measures to promote appropriate private investment, including facilitating the international flow of capital. Government investment may be needed in areas where market failure leads to under-investment—human capital and open access resources, for example. Human capital improvements for SARD may require pubic investments in rural health services, education, agricultural research and extension. Similarly, public investments may be needed in open access resources such as many fisheries and some forests, as well as in state-owned resources such as rural infrastructure. The usual tests of acceptable environmental impacts and social rate of return on the committed funds should be applied in appraising all such public investments.

The threat to SARD posed by population growth implies that, at least in the long run, population growth must be slowed. Instruments may include measures that improve the education of girls and the status and employment prospects of women, as well as public health measures to give women control of their own fertility. Population policy also extends to measures affecting the geographical distribution of people. Examples include measures to control rural-urban differentials that affect the rate of urban migration, and land settlement schemes to relocate people from over-populated to less densely settled areas.

Taxation and asset redistribution measures may be used to address the equity dimension of SARD. Policies may be adopted that specifically target the poor, such as famine relief programs, food for work schemes and provision of free or subsidized services for the poor. Striving for a reasonable standard of living for today's poor is important to enable poor parents to secure a more sustainable future for their children by providing them with a better start in life.

Policies relating to agricultural and rural development. Within the rural sector, rural infrastructure improvements can contribute to SARD by giving people better access to services and facilities that enhance the productivity of private rural capital. The proper maintenance of existing infrastructure is also important. Like all public investments, rural infrastructure developments need careful appraisal, including proper consideration of their environmental impacts.

Improvement of human capital in the rural sector is an often overlooked aspect of sustainability. Rural education and health services can help produce a rural population better able to manage agricultural resources. Investments in agricultural education, research and extension can promote SARD through the development and uptake of better production technologies. Through extension efforts it may be possible to change attitudes and values towards sustainability amongst the whole population.

Agricultural research policy and management will influence the efficacy and efficiency of R&D efforts directed towards SARD. The aim must be to focus efforts on issues that matter, and to make research responsive to the needs and circumstances of the intended beneficiaries.

In many less developed countries, the sustainable funding and staffing of agricultural research is a problem needing the attention of policy makers. The burden on the public sector may be eased, to some extent, by measures such as the creation and enforcement of patent rights that will encourage greater private agricultural R&D activity.

Policy interventions to date relating to agricultural prices arguably have mostly had negative impacts on SARD. In industrialized countries, setting prices above their equilibrium levels has led to agricultural surpluses, and has also led to inefficient and sometimes unsustainable levels of intensification. These policies have also often distorted international markets to the detriment of unsubsidized producers, most of whom are in the less developed countries. On the other hand, the tendency in less developed countries to take measures to hold down food prices has damaged the livelihoods of their own agricultural producers and may have obliged some poor farmers to degrade their resources.

Some threats to SARD come from climatic or other disasters, and policies to prevent or mitigate such events deserve consideration. So too do measures to help agriculturalists recover after such events have occurred. Foreign aid may be available to help rehabilitate devastated areas or industries in less developed countries.

It will sometimes be appropriate for governments to participate directly in agricultural production or marketing as part of the process of SARD. Recent trends are towards more reliance on the private sector. On the other hand, the fact that governments may have access to capital at lower rates of interest than private investors may justify government participation in long-term ventures such as forestry.

Especially in poorer countries, the goal of sustainable rural livelihoods may require policy interventions to help rural people escape from 'poverty traps'. Unfortunately, finding a package of policy instruments to do this is notoriously difficult, although some helpful measures are known.

Policy interventions relating to food and nutrition include the above-mentioned poverty alleviation measures to ease food security problems for the poor. In addition, overall food security may be enhanced by correcting any bias in agricultural research and extension programs in favour of cash crops at the expense of food crops. Pure food laws and dietary education can also help to improve nutritional standards, and hence human health and productivity.

Among policy options for SARD are various instruments relating to property rights. These include:

- reallocation of resource property rights between public, communal and private ownership;
- redistribution of privately owned resources among private individuals;
- regulation of the use of state, common, and open access property resources; and
- measures to encourage the efficient and sustainable use of resources.

All four types of policy interventions are likely to have impacts on sustainability, efficiency and equity. Moreover, where there is change of ownership, the full impacts of a particular policy may be difficult to predict. By their very nature, changes in property rights make some individuals better off and others worse off, meaning that such policies are often divisive.

Property rights may be a source of inefficiency if they do not provide for the transfer of the resources to those best able to use them. Inefficiencies will also follow if resource managers do not have reasonable security of title. For example, the lack of a transferable title to land may prevent the users of that land from gaining access to formal credit. Similarly, the lack of security of tenure may affect the care with which the land is conserved and used. In such situations, tenure reforms that are more consistent with SARD might be designed. Security and transferability of title can be of benefit for other forms of property, including rights of access to open access resources such as fisheries and to state property such as irrigation water.

Institutional development is important for SARD. In this context, 'institutions' are defined as the rules, conventions and other elements that form the structural framework of social interaction. Both community-based and market-based institutions need to be considered by policy makers, although there is usually more scope for them to influence market-based institutions than community-based ones.

SARD requires some fundamental changes in attitudes and values throughout society. Policies to promote a more 'conservation-minded' approach to agriculture may be promoted through the media, by working with 'change agents' of all types, and through local and national organizations, such as religious establishments, political parties or women's associations.

Policy attention to market-related institutions may be directed to making markets work better and to encouraging their development. That means removing unnecessary restrictions on market operations and establishing an environment in which entrepreneurship can prosper. Fair trading laws may need to be established or improved and some government assistance may be provided for the establishment of small business ventures.

Policies aimed at establishing democratic and participatory processes. Some SARD policies need to focus on establishing democratic and participatory processes. Decentralization may bring policy making about local matters closer to the people who will be most affected by those decisions. However, decentralization will not work unless sufficient funds and other resources are decentralized along with the responsibilities for action. Nor can a central government properly divest itself of those policy responsibilities that must be carried out in a coordinated way, including priority setting for regional activities.

People's participation is widely understood to be crucial for SARD. Unfortunately, getting participation to work is not easy. Suggested elements of policy to this end include the following:

- A truly participatory approach requires reasonable development of democratic processes, including respect for human rights, concern for the status of women, children and minorities, and reasonable standards of law and order and security.

- There is often a need to develop skills and to change attitudes and values both in local communities and among agricultural professionals; there may be a need to 'transform the learning environment', to change rural organizations into 'learning organizations'.

- Agricultural research and extension systems must be sensitive to the needs and circumstances of their clients.

- A policy of promoting SARD by working through local organizations, including NGOs, may circumvent some of the difficulties in transforming hidebound government organizations into people-sensitive entities.

- People's participation may be improved by making proper use of appropriate communication and information technology. The flow of information, in both directions, between the rural people and policy makers and planners at various levels needs to be as effective as possible. That may be attained using a variety of channels of communication, from local meetings and consultations, through to the use of modern electronic media, where appropriate.

Policies designed specifically to influence natural resource use. Policy approaches under this heading include the range of options of:

- direct government action;
- use of control instruments; and
- use of economic incentives.

Where, because of market failure or for other reasons, there is rural resource degradation or contamination of the environment, governments can act directly to try to correct these problems. For instance, governments can fund and undertake land conservation or rehabilitation works.

When direct government involvement in resource management is not appropriate, policy makers may turn to various types of control measures. These may include bans on certain activities or inputs, or the imposition of obligations on resource owners or managers. Such control measures are typically enforced via penalties for non-compliance. Inspectors have to be employed to detect those who break the rules. However, by the very nature of agriculture, enforcement is difficult and expensive, and may encourage corruption.

Because of these difficulties, economic incentives often offer better solutions to market failures affecting SARD than do control measures. Incentive approaches that have been suggested include:

- targeted subsidies
- tradeable resource shares
- individual transferable rights
- transferable development rights
- tradeable emission permits
- environmental taxes
- resource pricing
- effluent charges
- user charges
- deposit refund schemes
- environmental bonds.

All or most might be adapted to tackling particular resource conservation or environmental management problems relating to SARD. Unfortunately, experience in several countries suggests that the administrative and other impediments to the use of such measures can be considerable.

Policy programming

Good policy making for SARD entails an iterative, continuous cycle, with the following stages:

- diagnosis of problems and opportunities
- design of possible interventions
- setting the scope of programs or projects, i.e., determining the scale of operations
- impact assessment
- appraisal
- decision making
- implementation (action)
- monitoring and evaluation

then back to diagnosis. Not all steps may be followed on every occasion, and there may be looping back as ideas are developed and new information accumulated.

POLICY GUIDELINES FOR SARD

MAIN REPORT

1. INTRODUCTION

1.1 Historical background

One of the earliest expressions of concern about sustainability and food security was made by Malthus late in the eighteenth century. He assumed constant technology and geometric population growth, and so concluded that food shortages were inevitable. While for many years events appeared to prove Malthus wrong, pessimism was rekindled in 1972 in *The Limits to Growth* by Meadows et al. Their central dismal prophecy of a cataclysmic decline in population and production provoked a storm of controversy. Critics, among them many economists, argued that the prophecy was flawed because it failed to take proper account of the scope for substitution in production and consumption that would be induced, mainly through impacts on market prices, as resource availabilities and other circumstances changed. In particular, Meadows et al. were criticized for not accounting for the scope to substitute human capital in the shape of technological progress for other resources.

Notwithstanding these criticisms, both informed and popular concern for the environment continued to grow and led to the convening of a number of important meetings to discuss sustainability.

In 1987, the World Commission on Environment and Development was convened and did much to raise awareness about the need for and nature of sustainability (WCED 1987).

Reflecting growing concerns about the sustainability of agriculture, the FAO/Netherlands Conference on Agriculture and the Environment, was held in April 1991 at 's-Hertogenbosch (den Bosch), the Netherlands. This conference generated an important collection of papers (FAO 1991*a* to *v*) and led to a Declaration and Agenda for Action on Sustainable Agriculture and Rural Development (SARD hereafter). This Declaration identified three essential goals for the attainment of SARD:

- food security;

- rural employment and income generation (aimed at eradicating poverty); and

- natural resource conservation and environmental protection.

The Declaration noted that in many parts of the world these goals, and especially the eradication of poverty, are not easily attainable. It was pointed out that the achievement of SARD will require some fundamental changes and adjustments within agriculture based on consistent commitment to appropriate policies, backed by adequate resources.

Subsequently, the very influential United Nations Conference on Environment and Development (UNCED) was convened by the General Assembly of the United Nations. It took place in Rio de Janeiro, Brazil, from 4 to 14 June 1992. (This conference is also known as the Rio Conference or the Earth Summit.)

The Rio Declaration on Environment and Development contains 27 general principles to guide states and people about their rights and obligations in a 'new and equitable partnership' in matters of development and the environment. Agenda 21 of that Declaration is a comprehensive action plan for the period to the year 2000 and beyond. It contains 115 program areas relating to socio-economic aspects, natural resource conservation and management, the roles of major social groups, and the means of implementation. The chapters dealing with the agriculture sector (including forestry and fisheries) are:

- Chapter 10 (Integrated approach to the planning and management of land resources);

- Chapter 11 (Combating deforestation);

- Chapter 12 (Managing fragile ecosystems: combating desertification and drought);

- Chapter 13 (Managing fragile ecosystems: sustainable mountain development);

- Chapter 14 (Sustainable agriculture and rural development);

- Chapter 17 (Oceans and living marine resources); and

- Chapter 17 (Freshwater resources).

Drawing on the outcome of the den Bosch Conference, Agenda 21 articulated at the Rio Conference states that the major objective of SARD is to increase food production in a sustainable way and to enhance food security. It was stated that this will involve:

Box 1

SUSTAINABLE DEVELOPMENT AND GROWTH

Source: WCED 1987.

Sustainable development involves more than growth. It requires a change in the content of growth to make it less material- and energy-intensive and more equitable in its impact. These changes are required in all countries as part of a package of measures to maintain the stock of ecological capital, to improve the distribution of income, and to reduce the degree of vulnerability to economic crises.

- education initiatives, utilization of economic incentives and the development of appropriate and new technologies, thus ensuring stable supplies of nutritionally adequate food, access to those supplies by vulnerable groups, and production for markets;

- employment and income generation to alleviate poverty; and

- natural resource management and environmental protection (Agenda 21, ch. 14, para. 2).

In the same paragraph it is argued that, to create the conditions for SARD, major adjustments are needed in agricultural, environmental and macroeconomic policy, at both national and international levels, in developed as well as less developed countries.

1.2 The role of FAO in policy making

The main role of FAO in policy making, both in general and specifically in relation to SARD, is to respond to requests for advice and assistance from member countries. Such support is offered on the basis of an overall philosophy that is set out in the document *Outline for Guidelines for FAO Policy Assistance in the Agricultural Sector* (FAO n.d.). While these guidelines are still provisional, they suggest that the overall work of the Organization, and therefore the policy advisory work related to SARD, should be guided by the pursuit of the goals of **growth**, **equity**, **efficiency** and **sustainability**. In keeping with these goals, the general aim of FAO in offering assistance in policy analysis and formulation for the agricultural sectors of member countries is specified as devising, in partnership with governments and at their request, the practical means to bring about sustainable improvements in national economic, social and human welfare.

Converting such overall goals just mentioned into operational objectives to guide policy making entails judgement and leaves scope for differences of view and emphasis. Not surprisingly, therefore, there are different articulations of the objectives for agricultural and rural development generally and for SARD in particular. As set out in the provisional guidelines, however, in assisting member governments to make policy choices, FAO should seek to help them to:

- clarify national sectoral goals;

- distinguish between policy means, as a way of meeting goals, and policy ends, i.e., the goals themselves;

- propose policy options and program alternatives by which governments can meet selected goals;

- evaluate and analyse the potential effects of these alternatives quantifying any tradeoffs;

- design action programs which include institutional strengthening or human resource development plans intended to implement the chosen alternative; and

- devise systems to monitor progress and measure the impact of policy change.

Clearly, these procedures are fully consistent with a commitment by FAO to promote SARD in accord with the recommendations of UNCED.

Before proceding further, it is important to comment on specific aspects of the approach used in this report. Some readers may question the definition of SARD that forms the basis for this document. For instance, it can be argued that growth should include efficiency, and hence, the latter does not need to be a separate criterion for SARD. It can be noted that equity is not necessarily the same thing as poverty reduction, or that the concept of sustainability (following the Bruntland Report [WCED, 1987]) could be reserved to inter-generational equity. While acknowledging these caveats, this report adheres to the definition of SARD as espoused by FAO and other international organizations. Certainly, the definition of SARD is not static, but one takes the widely accepted concept at any point in time as the one upon which to build a report.

1.3 Scope of this report

The terms of reference set for this report, given in the Preface, refer to the formulation and assessment of policies in the agricultural and rural sector. (For this purpose, agriculture is understood to encompass fisheries, forestry, and, where relevant, associated up- and down-stream activities.) This breadth of coverage makes sense because agriculture, forestry and fisheries are critically important in the follow-up action to UNCED. Few if any other sectors are affected as much by the sometimes conflicting demands of development and environmental protection. Moreover, by the nature of the activities they are engaged in, agricultural producers, including foresters and fisherfolk, are the managers of a significant share of the world's natural resources.

The agricultural and fisheries sectors also fulfil the vital roles of supplying food essential for human survival, and all three sectors supply industry with raw materials. Large numbers of people depend for their livelihoods on these sectors, especially in the less developed countries. Striking the right balance between economic, social and environmental goals is crucial for the welfare of these people and their families. Most of the poverty in the world is found in rural areas, so the goal of the elimination of poverty requires that special attention be given to raising productive employment in these areas. In those places where the natural resource base is too fragile to withstand further intensification, that means developing rural value-adding industries to provide the required employment opportunities.

1.4 Levels of policy intervention

Within the broad agriculture and rural sector, it is possible to analyse policy options at various levels of aggregation:

- At the global level, dealing with the sustainability of world agriculture as a whole. A key issue at this level is the growth of demand for food due to population growth and increased consumption per head of some foods due to rising incomes. Consideration of this issue focuses attention on the scope and means that exist to expand overall agricultural production in tune with the expanding demand in the face of resource degradation, possible deleterious climatic change and an uncertain future supply of ever more productive technologies.

- At the intermediate level—the regional, national, district or area level. Here, concern within the agricultural sector is focussed on the sustainability of particular agro-ecosystems, including forestry or fisheries systems, and value-adding related rural industries.

- At the local level—at the level of a field, a farm-household, a rural business, or a local community. Issues include the sustainability of the way of life of the people living in these systems, as well as the changes that may be taking place in the quantity and quality of resources to which these people have access.

At this latter, most disaggregated level, the agricultural sector is uniquely characterized by very many production units operating in diverse natural and socioeconomic environments. It is therefore not possible to prescribe a single strategy for sustainability that will be widely applicable. It is also impossible to define specific methods of management that will enable large groups of agricultural producers to operate in a sustainable way. Indeed, the very diversity of agricultural production systems is part of their strength, since it reflects adaptations to different needs and circumstances, including access to resources and markets. Without such diversity, agriculture would be less sustainable.

Clearly, any policy guidelines for SARD must account for the nature and diversity of local systems. Unless policies are in tune with these local conditions, they will not work. That might seem to make the task of setting out policy guidelines too daunting. Yet there are sufficient commonalities among many of these diverse systems to permit some useful generalizations about the policies that can help promote SARD. Some of these policy guidelines, such as those relating to macro-economic settings or to various kinds of market failure, apply to most sectors of an economy. Others are more specific to agriculture, such as those relating to rural infrastructure or to the strengthening of human capital applied in the agricultural sector.

In what follows, attention will be given to the development of policy guidelines at all levels, although the focus will be principally on:

- sectoral polices at national level; and

- measures that can be taken to promote SARD at the local level.

At the end of the report, some comments are made about international aspects of policy making for SARD.

2. THE CONCEPT OF SARD

2.1 Definitions

The concept of sustainable development is an evolving one, and there are many definitions in the literature, some very similar, and others markedly different. Pezzey (1992) lists 27 definitions of sustainability and sustainable development, and there are many more.

According to the Bruntland Report (WCED 1987):

> *Sustainable development* is development that meets the needs of the present without compromising the ability of future generations to meet their own needs.

In 1988, and on the basis of the Bruntland Commission definition of sustainable development, the FAO Council defined SARD as:

> ... the management and conservation of the natural resource base, and the orientation of technological and institutional change so as to ensure the attainment and continued satisfaction of human needs for present and future generations. Such sustainable development (in the agriculture, forestry and fisheries sectors) conserves land, water, plant and animal genetic resources, is environmentally non-degrading, technically appropriate, economically viable and socially acceptable (FAO 1989).

Both of these definitions are strongly anthropocentric—the focus is on the sustainable welfare of humans. They contrast with definitions of sustainability proposed by some ecologists. As Schuh and Archibald (1996, p. 3) note, it makes sense to concentrate on the welfare of people since any operational approach to the conservation of natural ecosystems must be rooted in the beliefs and values of society. Of course, those beliefs and values may embrace ecological and environmental concerns.

2.2 Alternative views

Notwithstanding some wide acceptance of the UNCED and FAO definitions of sustainable development and SARD, respectively, there is still plenty of disagreement about the definitions themselves and about their interpretation. Problems in finding an operational definition of sustainability are the main reason for some economists, such as Beckerman (1992), arguing that sustainability is a meaningless notion. According to Pretty (1994, p. 39):

> ... any attempt precisely to define sustainability is flawed. It represents neither a fixed set of practices or technologies, nor a model to describe or impose on the world. The question of defining what we are trying to achieve is part of the problem, as each individual has different objectives. ... Sustainable agriculture is ... not so much about a specific farming strategy as it is a systems-oriented approach to understanding complex ecological, social and environmental interactions in rural areas.

In other words, Pretty was arguing that SARD is a learning process, not a goal—a view also held by some ecologists. Others have argued that sustainable development is about sustainable human livelihoods, particularly the relief of poverty now and in the future, implying a strong focus on current and inter-generation equity.

While the views of sustainability of people from different disciplinary backgrounds are often quite different, they are best seen as complementary, not conflicting. The three legs of the sustainability 'tripod' can be viewed as representing the economic, ecological and sociological schools of thought. Without all three legs the tripod will not stand. Each leg gives support to the others. Only if all three are

---

Box 2

SARD AS A PROCESS RATHER THAN A PRODUCT

Adapted from Carley, 1994, p. 2

The human element in sustainable development can be seen as a continuing process of management and mediation among social, economic and biophysical needs that results in positive socio-economic change that does not undermine the ecological and social systems upon which communities and societies are dependent. Its successful implementation requires integrated policy, planning and social learning processes; its political viability depends on the full support of the people it affects through their governments, social institutions and private activities linked together in participative action.

From this perspective, the process of sustainable development will always precede the product—SARD is a journey rather than a destination.

---

firmly on the ground can the whole entity be strong enough to use.

2.3 The position of FAO

The FAO definition of SARD has been given above. FAO criteria for SARD are (Pétry 1995a, p. 2.3):

- Meeting the basic nutritional requirements of present and future generations, qualitatively and quantitatively, while providing a number of other agricultural products.

- Providing durable employment, sufficient income, and decent living and working conditions for all those engaged in agricultural production.

- Maintaining and, where possible, enhancing the productive capacity of the natural resource base as a whole, and the regenerative capacity of renewable resources, without disrupting the functioning of basic ecological cycles and natural balances, destroying the socio-cultural attributes of rural communities, or causing contamination of the environment.

- reducing the vulnerability of the agricultural sector to adverse natural and socio-economic factors and other risks, and strengthening self-reliance.

These four criteria might be given the shorthand names of **food security**, **equity**, **responsibility** (in resource use and environmental management) and **resilience**.

What is not explicitly mentioned in this list of criteria, although it is certainly implied, is **growth**. In the face of growing populations and rising human aspirations, agriculture must expand. It must do so not only to feed the growing numbers of people, but also to contribute to the reduction in poverty and deprivation by providing affordable food for the poor and productive employment for rural people. This is not to suggest that growth, by itself, is enough, but, without growth, the other criteria cannot be met. At the same time, it is clear that the growth must occur in a sustainable way, without the negative impacts on resources, the environment and income distribution that have too often happened in the past. This is surely the central challenge for SARD policy makers and planners.

2.4 Issues of space and time

2.4.1 Systems and sub-systems

As Lynam and Herdt (1989) have pointed out, 'sustainability is first defined at the highest system level and then proceeds downwards; and as a corollary, the sustainability of a system is not necessarily dependent on the sustainability of all its sub-systems.' Sub-systems may be defined in terms of scope or scale. Thus, at the highest level, what matters is the sustainability of the planet earth. Invoking the Lynam and Herdt principle implies that individual sub-systems need not all be sustainable for global sustainability to be achieved, although the requirement of global sustainability constrains the tolerable levels of unsustainability of sub-systems.

Within agriculture, clearly not all agricultural sub-systems may be sustainable. In some cases, the prevailing system may be causing a progressive depletion of the natural resource base, for example, through soil erosion. In places where agriculture was traditionally based on shifting cultivation with bush fallow, population growth and perhaps the loss of access to some land may have led to an unsustainable reduction in the fallow length. Sometimes, in such cases, it may be possible for the system to evolve, perhaps with improved and appropriate technologies, into a sustainable form of settled agriculture. In other cases, where the natural environment is too fragile, there may be no technologies consistent with preventing the degradation of the resource base and providing all the people with sustainable livelihoods.

7

Box 3

APPROACHES TO SUSTAINABLE DEVELOPMENT

Adapted from Munasinghe and Cruz 1995, pp, 8-9

The concept of sustainable development has evolved to encompass three major points of view: economic, social, and environmental, as shown in the figure (Munasinghe 1993)

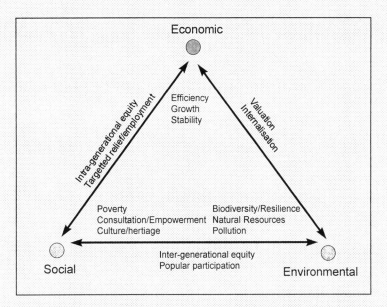

The *economic* approach to sustainability is based on the concept of the maximum flow of income that could be generated while at least maintaining the stock of assets (or capital) which yields those benefits. Notions of optimality and economic efficiency underlie the allocation and use of the scarce resources. Questions arise about what kinds of capital need to be maintained (e.g., natural, manufactured and human capital) and their substitutability. There are also difficulties in valuing these assets, particularly ecological resources. The issues of uncertainty, irreversibility and catastrophic collapse pose additional difficulties.

The *social* concept of sustainability is people-oriented, and relates to the maintenance of the stability of social and cultural systems, including the reduction of destructive conflicts. Equity is an important consideration from this perspective. Preservation of cultural diversity and cultural capital across the globe, and the better use of knowledge concerning sustainable practices embedded in less dominant cultures, are seen as desirable. There is a perceived need for modern society to encourage and incorporate pluralism and grass-roots participation into a more effective decision-making framework for socially sustainable development.

The *environmental* view of sustainable development focuses on the stability of biological and physical systems. Of particular importance is the viability of subsystems that are critical to the global stability of the overall ecosystem. Furthermore, 'natural' systems and habitats may be interpreted broadly to include man-made environments such as cities. The emphasis is on preserving the resilience and dynamic ability of such systems to adapt to change, rather than conservation of some 'ideal' static state. Natural resource degradation, pollution and loss of biodiversity reduce system resilience. (*continued*)

Such 'poverty traps' are all too common in various parts of the world. Sometimes it will be possible to find policy interventions to allow people to break out of such traps. Examples include the introduction of improved technologies, perhaps combined with asset redistribution, or the development of associated rural industries. On other occasions, no such answers may be found that could be implemented at reasonable cost, and it will be necessary for the excess population to move to other, less threatened areas. Resettlement schemes have been used to ease population pressure on land in a number of countries, such as Indonesia (although not always with the intended beneficial social and environmental benefits). While it is true that not all agricultural systems may be sustainable, now or prospectively, it is also the case, as indicated by Lynam and Herdt, that not all systems *need* to be sustainable. Some substitution of resources may be desirable, even necessary, for overall sustainability to be attained. For example, for a less developed country with relatively good forestry resources to impose on itself the condition that the harvest rate should be no higher than the regeneration rate may be a mistake. It would exclude the possibility of converting at least some forest areas to more productive agricultural use in crop or livestock farming. Some of the loudest calls for countries in the South to conserve their forests come from the North where large areas of forest and woodland were successfully converted to other uses long ago. It seems unreasonable to expect poorer less developed countries to deny themselves the same opportunities.

Depleting one natural resource may be consistent with the goal of sustainability in general, and SARD in particular, provided that the total stock of capital is not depleted. This might mean, say, investing in human capital to compensate for the unavoidable or considered depletion of capital inherent in a forest, a fishery or farmland. Much of the debate between economists and environmentalists about what constitutes sustainability centres on different perceptions of the extent to which such substitution is possible. Economists generally take an optimistic view of substitution possibilities and environmentalists a pessimistic one.

The need to take a global view of sustainability is also implied by the growing linkages between countries. With increase in international trade and greater mobility of people, capital and technology, it is no longer necessary for countries to aim for self-sufficiency. Indeed, to do so is to deny access to the advantages of international trade through exploiting comparative advantage (see Box 5).

2.4.2 The time dimension of SARD

The FAO definition of SARD refers to the needs of present and future generations. The time horizon for planning and policy making implicit in such a definition is very long. However, because the long-term future is so hard to foresee, there are severe operational difficulties in setting a very long time horizon for policy making and planning.

Clearly, the choice of the appropriate time scale will depend on circumstances. Longer horizons may be unavoidable for long-term processes such as forestry while in other cases the here-and-now urgency of some environmental threat may need to be dealt with on a short-term basis. It is also important to distinguish between the time period for which plans are laid, herein called the planning horizon, and the time over which the consequences of those plans may be experienced. Planning horizons of ten or fifteen years are usually about as long as it is plausible to consider. Even over such periods, the ability to account

Box 4

SUSTAINABLE LIVELIHOODS AS AN INTEGRATING CONCEPT

Drawn from Chambers and Conway 1992, pp. 5-8.

In calling for a new analysis, the Advisory Panel of the World Commission on Environment and Development proposed sustainable livelihood security as an integrating concept, and made it a central part of its report (WCED 1987, 2-5). The definition was:

Livelihood is defined as adequate stocks and flows of food and cash to meet basic needs. Security refers to secure ownership of, or access to, resources and income-earning activities, including reserves and assets to offset risk, ease shocks and meet contingencies. Sustainable refers to the maintenance or enhancement of resource productivity on a long-term basis. A household may be enabled to gain sustainable livelihood security in many ways—through ownership of land, livestock or trees; rights to grazing, fishing, hunting or gathering; through stable employment with adequate remuneration; or through varied repertoires of activities.

The Panel argued that this was an integrating concept since sustainable livelihood security was a pre-condition for stable human population, a prerequisite for good husbandry and sustainable management, and a means of reversing or restraining destabilizing processes, especially rural to urban migration. Sustainable livelihoods were seen as a means of serving the objectives of both equity and sustainability.

Chambers and Conway argue that sustainable livelihoods also provide the resources and conditions for the enhancement and exercise of capabilities, by which they mean the ability to perform certain functionings, basic to what a person is capable of doing and being. It includes, for example, to be adequately nourished, comfortably clothed, to lead a life without shame, to be able to visit and entertain one's friends. The word capability thus has a wide span, with different specific meanings for different people in different situations. Using this concept, they propose the following working definition of sustainable livelihoods:

A livelihood comprises the capabilities, assets (stores, resources, claims and access) and activities required for a means of living; a livelihood is sustainable which can cope with and recover from stress and shocks, maintain or enhance its capabilities and assets, and provide sustainable livelihood opportunities for the next generation; and which contributes net benefits to other livelihoods at the local and global levels in the short and long term.

for changes in the technological, social, political and economic situation some several years into the future is very limited. In these circumstances, therefore, it is necessary to lay plans and set policies on a shorter term basis while giving due weight to the implications of today's actions for future generations. This requires the adoption of a **sustainability principle** in policy making and planning.

Adoption of a sustainability principle means making sure that actions taken now do not unduly limit the opportunities for people in the future to live as well or better than we live today. It means that we should not deplete the total resource stock that we pass on to the next generation. In fact, short of some massive

calamity, it will take many years before global population growth can be halted and eventually reversed. Therefore we need to pass on the planet to the next more numerous generation in a *more* productive condition than it was when we inherited it.

The value judgements implicit in the ideas presented in the previous paragraph do not mean that we must pass on better stocks of all resources. To do so with the exhaustible resources such as fossil fuels is clearly impossible. Rather it is the total bundle of resources that needs to be improved from generation to generation. As one resource stock is run down, the total bundle can improve only if there are at least compensating improvements in the

---

Box 5

THE NEED FOR A GLOBAL PERSPECTIVE

From Schuh and Archibald 1996, p. 4

There are a number of reasons for taking [a global or international perspective to the sustainability issue]. First, it is consistent with FAO's global mandate. Second, today's economy is truly globally interdependent, with all national economies linked to the international economy in one way or another. Global sustainability is thus the critical issue. Third, countries that fail to meet the conditions for global efficiency implied by this perspective sacrifice national income for their citizens and thus fail to meet the goal of sustainable development. In effect, addressing the issues of sustainability at the project level, which is a common approach, without paying attention to broader efficiency criteria, is not consistent with the theory of second best (*i.e. the usual border price criteria for efficiency is in fact a second best condition when one views the global economy -- editor's note*).

---

stocks of some other resources that can substitute for what is lost. In particular, human capital, embodied in such things as improved technologies and better management skills, can substitute, at least to some extent, for depletion of some natural resources. As noted above, there are differences of opinion about the extent to which such substitution will be possible in the future.

The concern for the welfare of future generations embedded in the SARD definition may appear to be in conflict with the practice of using a positive discount rate to bring future costs and benefits to present values when planning investments such as agricultural projects. Positive discount rates may seem to discriminate against future generations because projects with costs in the distant future but benefits in the near term will be favoured. On the other hand, investments that have long-deferred benefits will not appear attractive with discounting.

The negative impact on sustainability may be particularly strong for countries and individual resource managers who face high costs of capital. Heavily indebted nations may have limited ability to borrow internationally so that the actual or opportunity cost of capital is likely to be higher than less-indebted countries. Similarly, the productivity of capital for poor farmers in less developed countries may be very high because such people are often unable to access formal markets for credit. High interest rates may mean that long-term investments, such as land improvements, cannot be afforded.

Such reasoning has led some environmentalists to argue for a zero or negative discount rate to

promote sustainability. However, it is not clear how actual interest rates could be forced down to zero without savage and politically unrealistic cuts in current levels of consumption. Nor is it clear that arbitrary use of a zero discount rate for public project appraisal makes sense. It would imply accepting projects showing low returns on investment when others would be available showing positive rates of return. To waste productive investment opportunities in this way would be to deny future generations the benefits from more productive uses of scarce capital. Moreover, lowering the discount rate may also lead to the adoption of investments that are resource degrading, such as logging low-yielding or difficult areas of natural forest, or fishing already over-exploited fisheries.

The consensus of opinion is that attempts to attain sustainable development by manipulating the interest rate are misplaced. It would be desirable to solve the problems of the excessive levels of indebtedness of some countries. Similarly, there is merit in promoting the development of financial systems to make it possible to deliver credit to rural resource managers who need it at reasonable cost. But tinkering with the discount rate to try to attain inter-generation equity is inappropriate—as is the not unheard of practice of government ministries deliberately selecting low discount rates in order to put pet projects in a better light. A more satisfactory approach is to apply conventional investment criteria to all projects using normal discount rates, but also to invoke the sustainability principle in the analysis of public investment programs. That means selecting projects on the basis of discounted costs and benefits subject to the condition that the overall stock of capital (natural, artificial and human) across all investments is enhanced for future generations (Markandya and Pearce 1991).

While the implementation of such a rule may be easier said than done, at least until much improved methods of measuring and monitoring changes in capital stocks are developed and brought into use, it is an important guiding principle for policy makers. For the time being, it might be implemented using a multi-criterion decision rule in which the return to the investment is considered along with a number of operational aspects. These may include the imposition of safe minimum standards in environmental protection and limits on the rates of depletion of certain key resources for which substitution options are limited. Also, when natural capital such as mineral reserves or forests are depleted, it will be wise to make certain that other compensating investments are made.

3. THE CHALLENGE: THREATS TO SARD

For thousands of years, agriculture, including forestry and fishing, has provided the food and other materials necessary for human survival and well being. Yet today, as we have seen, there is growing concern about the capacity of agriculture to meet the needs of future generations. The challenge to SARD comes from population growth, the increasing demand for food, the need to reduce poverty, and the threat to agricultural resources of land, water, and living things.

3.1 Population growth

By the year 2000, world population is expected to be about 6.25 thousand million, compared with 3.0 thousand million in 1960. About 80 per cent of the year 2000 population will live in the less developed regions. While the global rate of population growth has slowed to less than 1.7 per cent annually, the absolute increase, of nearly 100 million annually is more than it has ever been. Even if birth rates are brought down relatively quickly, the high proportion of young people in those countries where population growth is still high means that the momentum for growth will continue at least until the middle of the next century.

The rates of population growth are very uneven, with most industrialized countries experiencing close to zero rates of natural increase, and most of the growth coming in the less developed countries. In some parts of the world, as in much of sub-Saharan Africa, population growth has tended to outstrip growth in food production. It is in just these less developed parts of the world that future food production needs to be increased more rapidly in order to improve food security, especially for the poor.

3.2 The growing demand for food and fibre

Global demand for food is likely at least to double by 2030. Most of this increase will occur in the less developed countries of Africa, Asia and Latin America. This concentration in part reflects the rapid population growth in these areas. It also reflects, all being well, desirable improvements in the quantity and quality of food consumed per head as incomes rise (McCalla 1994). In sub-Saharan Africa, to feed the increasing numbers and to reduce presently unacceptable levels of food insecurity require a more than tripling of agricultural production by 2025 from a 1990 base (Crosson and Anderson 1995a, 2-4).

In addition to the population growth, changes will occur in the pattern of demand. In many of the newly industrializing countries, and in those less developed countries that have been successful in promoting economic growth, rising incomes are likely to lead to significant increases in demand for meat and other animal products (including fish). Such changes in demand for animal protein are unlikely to be fully or even mainly met from livestock grazed on land not suited for cropping. They therefore imply an increased demand for grain to be used as feed for intensively kept livestock and farmed fish. To produce one kilogram of human food as intensively reared animal protein takes several kilograms of grain or other stockfeed. If there is not a compensating increase in grain production, the diversion of these feeds to animal production is likely to drive up prices, reducing the access of the poor to staple foods.

The increased food production to meet the expanded demand will have to come mostly from increased yields. The reason is that there is little new land to be brought into production, and agricultural land is being lost to urban development, degradation and for other reasons. During the 1970s and 1980s, arable and permanent crop area per capita fell from 0.37 ha to 0.27 ha for the world as a whole. The fall in the less developed countries was sharper from 0.27 ha to 0.19 ha (Pétry 1995a, 1.3-1.4). These trends in arable area are likely to continue. They mean that the projected doubling in the demand for food by 2030 will be met only if yields per hectare are more than doubled.

At best, this is a daunting challenge. There are some who suggest that it is not attainable, or at least, not without seriously damaging consequences for the environment as a result of the required increases in agricultural intensity based on increased use of chemical inputs.

3.3 The challenges to reduce poverty

According to IFPRI (1995, p. 10), over 1.1 thousand million people in the developing world live in absolute poverty, with incomes per person of a dollar a day or less. Moreover, IFPRI (p. 8) estimates that about 800 million of these, or one in five of the population of the developing world, lack the economic and physical access to the food required to lead healthy and productive lives.

There is an urgent challenge to agricultural policy makers to find ways of reducing the scale of this deprivation, above all for humanitarian reasons. In addition, failure to tackle the problem quickly and effectively is likely to mean that the rural poor will be driven to migrate to the cities in search of livelihood. The swelling slums that surround many of the major urban centres in less developed countries present their own threats to the sustainability of social and political systems.

Poverty persists because people lack entitlements that they can use to earn an acceptable living. They may have no access to land, or to water to irrigate arid land, or to technologies appropriate to their needs and circumstances, or to markets that offer reasonable prices for what they can produce, or to employment opportunities that offer a reasonable reward for their labour. These problems of lack of entitlement are not easy to solve, yet finding solutions is fundamental to the pursuit of SARD. Unless policies for SARD lead to a reduction in poverty, they cannot be regarded as successful.

At least in some situations, there is a vicious circle linking poverty and unsustainability. The poor may be driven by their very poverty to degrade resources in order simply to survive (von Braun 1992). Certainly, they are likely to lack the capital and access to credit to invest in resource improvements such as land conservation measures. Moreover, as the resource base declines, at least per person, the pressures on the poor that force them into unsustainable ways may increase. In such cases, policy makers need to find a means to break the downward spiral. For example, a solution might lie in the provision of a 'food for work' program, with the labour used for conservation measures. Yet too much should not be made of the contribution of the poor to resource degradation, since it is the rich who are the largest consumers of global resources such as energy. Moreover, the kinds of resource degradation caused by the rich (global warming, pesticide run-off) tend to be different from the damage done by the poor (deforestation, erosion). Clearly, attention needs to be given to ways of changing the behaviour of both rich and poor to address these different issues if sustainability is to be attained.

3.4 Resource depletion and degradation

3.4.1 Land, water and the environment

Hard data on degradation of land and water resources on a world scale are lacking. There is much concern, no doubt much of it well justified, about losses due to soil erosion, salinization or desertification. Yet in a recent review of a number of studies, Crosson and Anderson (1995b) note that the historical loss of productivity due to land degradation, although not unimportant, has been rather low relative to the productivity increase due to other factors. Even under rather pessimistic assumptions, they estimate that the annual average rate of productivity loss due to land degradation in the less developed countries has been less than 0.2 per cent a year over the past 45 years. There is also concern about competition for water between agricultural, industrial and urban uses and about declines in water quality for agricultural and other purposes. However, Crosson and Anderson further conclude that the average rate of productivity loss on irrigated land due to degradation of water resources for irrigation has been no more than 0.3 per cent a year over the past 30 years.

While the above rates are far from insignificant in the long run, they do suggest that the scope for attaining SARD is likely to lie mainly in expanding the use of knowledge-intensive inputs in agriculture, rather than in giving absolute priority to preventing land and water degradation, or to efforts made with too little regard to costs to reverse past unfavourable trends in quality of land and irrigation water resources.

The shift in technology required to produce more food and fibre from less land has thrown into doubt the capacity of the natural environment to absorb the resulting pollution and contamination. Environmentalists emphasize the role of the environment not only as a source of resources or inputs, such as land and water, used in production, but also as a sink to absorb unwanted by-products of production and consumption, and as a direct source of amenities. They are concerned that this absorptive capacity is being overloaded and will be much more seriously threatened if present methods of agricultural production are greatly intensified. Apart from the perceived risk of a breakdown in the life-sustaining ecosystem itself, there is the lesser but still serious threat to human health as more pollutants from intensification are released into the environment. The challenge, therefore, is to find more intensive methods of production that impose a lesser burden on the environment. Pretty (1995) has argued that there are resource-conserving technologies, local institutional structures and enabling external institutions that meet this challenge. The task for policy makers is to provide the conditions for such positive changes to occur.

3.4.2 Loss of animal and plant species

The development and intensification of agriculture have led to loss of diversity in both domesticated and wild biological species, and the decline is continuing. In the past, humans have used well over 100 000 edible plant species and numerous animal species. Today, most of the world's population depends on only 12 crop species for their food staples. While the actual degree of concentration is overstated by these figures, in that many of the species used in the past were never very important, many scientists see the implied loss of diversity as worrying.

While some loss of biodiversity is unavoidable, and while there are certainly advantages from concentrating on the more productive species, there are grounds for anxiety that the decline may be socially and ecologically excessive. The erosion of the genetic resource base may mean the permanent loss of actually or potentially useful species or traits. Moreover, production systems are exposed to increased risks by relying on a narrow genetic base for domesticated species. SARD clearly requires that the downward trend in biodiversity be arrested.

3.5 Challenges to sustainable forestry

Deforestation is occurring at a rapid rate in many parts of the world. According to the *1990 Forest Resources Assessment* (FAO 1995*c*), the annual loss of tropical forests has been about 15.4 million ha, mostly in the less developed countries. Afforestation and reforestation in the tropics have been 1.8 million ha per year, representing only 12 per cent of deforestation.

In many of the less developed countries, deforestation is occurring as forest is cleared for farmland and rangeland to support growing human populations. The farming systems introduced on the cleared lands are not always sustainable.

While there has been an expansion of the temperate forest resource in the industrialized countries, the demand for forest products is expanding. There is also an increased demand for the services forests provide in protecting the land and water base, providing sustainable livelihoods for forest-dependent communities and conserving ecosystems and biodiversity (FAO 1993; 1994*b*).

3.6 Challenges to sustainable fisheries

Despite the rapid development of aquaculture, fishing remains largely a harvesting activity characterized by lack of property and user rights. The result is that many stocks and fishing areas are over-fished. According to one estimate, all the world's 17 main fishing grounds are being fished at or above their sustainable limits (O'Riordan, in FAO 1995*b*, 17-19). Some 70 per cent of the world's fish stocks have been classified as fully exploited, over-exploited, depleted or recovering from over-exploitation. Moreover, the functioning of these ecosystems and their capacity to recover from over-exploitation is uncertain. Management of fisheries is not easy and demands cooperation at several levels, from local to international. FAO has developed a

---

Box 6

SUSTAINABILITY OF HUMAN CAPITAL

From Schuh and Archibald 1996, p. 22

As we look to future decades, the fundamental issue of sustainability may be associated with issues surrounding human capital. There are a number of important issues here. The first is that the easily added accretions to new knowledge may already be behind us. This is reflected in the increasing costs of research in universities and research institutes, especially in terms of the increasing costs of equipment and instrumentation. Second, in the case of biological research for agriculture, there is some evidence that an ever larger share of the total research budget must go just for maintenance purposes. In other words, an even larger share of the research budget has to go just to sustain the gains realized in the past as micro-organisms and insects develop resistance to past treatments. Finally, research is for the most part a pure service sector of the economy. It has traditionally been difficult to raise productivity in the delivery of pure services. This is another reason why costs of doing research, and of producing a given accretion to our stock of knowledge, may rise over time.

It is true that there have been improvements in the technology of doing research. The various components of biotechnology are an excellent example. Moreover, we know very little in a systematic way about the technology of doing research or about the costs of doing it. As the share of increases in output accounted for by investments in science and technology grows, these issues will become increasingly important in the sustainability debate.

Code of Conduct for Responsible Fisheries. The Code places emphasis on the adoption of selective and environmentally safe fishing practices designed to maintain biological diversity, and safeguard aquatic ecosystems.

3.7 The challenge in developing human capital

Much of the discussion about SARD focuses on the physical resources used in agriculture. Yet it is clear that substantial increases in productivity will be needed to meet future demand for food. These increases must come from improved technologies, most of which will require large investments in education, research and technology development. The challenge, therefore, is to make sure that investments in these forms of human capital are sustained at the necessary rates. This challenge is critical because it seems that the easiest technological gains may have already been attained, so that a higher rate of investment will be needed in future than in the past to sustain the required rate of productivity increase (see Box 6).

4. THE APPROPRIATE POLICY FRAMEWORK FOR SARD

There needs to be an appropriate framework in place for the formulation and implementation of policies for SARD. Moreover, policy making for SARD needs to be approached accounting for some important features of the task. Elements that are likely to be important include:

- Taking a long-term, global perspective.

- Implementing complementary sectoral and macro-economic policies.

- Developing coordinated and consistent policies within the agricultural sector.

- Having an appropriate legal and institutional structure in place.

- Following an approach in which the inherent uncertainties of planning for SARD are recognized and accommodated.

Taking an adaptive, inter-disciplinary and multi-disciplinary 'systems approach' to policy planning. Each of these aspects is discussed further below.

4.1 The need for a long-term global perspective.

The need to take a long-term view in considering sustainability should not need emphasizing, yet there appears to have been a tendency in recent policy work to concentrate on the short term, perhaps driven by political expediency or short-term difficulties in macroeconomic management. A focus on SARD requires decision makers to take a more far-sighted view of their responsibilities.

The case for a global perspective rests on the point noted earlier that sustainability needs to be assessed at the highest level, and that sustainability of the global system does not require that every sub-system be sustainable. Technological changes in transport, communications and information science are creating a global economy in which goods, services and resources are widely traded. Countries, districts within countries, and even individual businesses and households, are increasingly able to benefit from trading in the progressively more integrated global market, and so need to give less attention to self-sufficiency. The principle of comparative advantage suggests that freer trade should facilitate increased and sustainable production from the available resources. It implies that countries can import resources that are in short supply, or can export resources that are in relatively abundant supply. There have even been suggestions that countries can export 'pollution' to where it can be better dealt with—a proposition that, not surprisingly, can excite considerable opposition!

The increasingly integrated global market should also mean that prices better reflect the relative global scarcities of resources, goods and services, sending more appropriate signals about these relative scarcities to producers and consumers. This should promote more sustainable behaviour, provided only that governments do not excessively distort domestic prices through inappropriate protectionist or taxation measures. According to Schuh and Archibald (1996):

> The significance of these develop-ments to the sustainable development issue is that this increased integration through international markets creates the potential for large efficiency gains from the international division of labor, increased specialization in production, and the realization of comparative advantage. The efficiencies from these processes will release the resources essential for realizing sustainable economic development on a scale not realized in the past.

Extending this line of reasoning further, some argue that trade liberalization will in fact promote sustainable development only if other problems, such as other market failures and inequity, are tackled at the same time. These critics argue that, in a world of rich and poor, it is too easy for the rich to use their trading power to the disadvantage of the poor. Certainly, the process of structural adjustment in a country as trade barriers and other regulations are dismantled often appears to exacerbate rather than cure problems of poverty and inequality, at least in the short to medium term. It may therefore be important for policy makers to give careful thought to the appropriate pace of change and for them to put in place safety net measures for those who will be most disadvantaged.

4.2 Harmonizing general and sectoral policies

A summary of some of the complex linkages between macroeconomic policies and the environment is provided in Figure 1, which is taken from Markandya (1994).

Macroeconomic policies can be as important, or more important, than sectoral policies in affecting resource use, income distribution and growth, and hence SARD. It is therefore vital that macro-level and sectoral policies are harmonized. In the less developed countries it has been common to put in place policies that in effect discriminate against agriculture; in the industrialized countries, the reverse is the case. Both types of policies tend to create serious inefficiencies in the use of resources, including agricultural resources, and therefore impede progress towards SARD.

4.3 Integrating action within the agricultural sector

It is similarly important to harmonize policies within the agricultural sector. This might not be as simple a task as it may seem, since policies affecting agriculture are not made solely within the agricultural ministry. So, improvements to rural transportation and infra-structure, and to rural health and education, are all vital components of efforts to progress towards SARD. Yet these areas usually fall within the responsibility of other ministries, not the ministry of agriculture. Unless systems are put in place to encourage harmonization of efforts, and to avoid conflicts between differ-ent groups and agencies, efforts may be dissi-pated.

Figure 1: Macroeconomic policies and environmental impacts

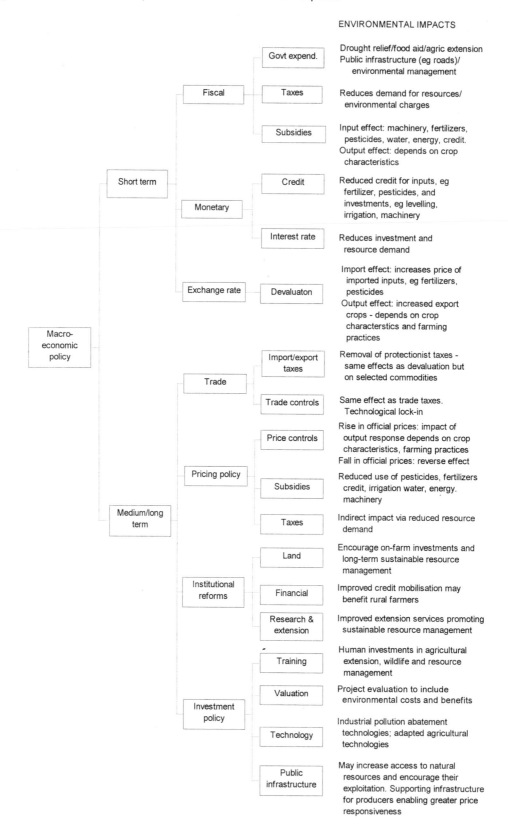

Source: Markandya (1994).

4.4 Building an appropriate legal and institutional structure

Market failure often occurs when the legal structure is too weak for people to enforce their property rights, or for governments to enforce regulations for the control of resource use.

Private property rights usually exist in resources or in commodities. Well-functioning markets can contribute to SARD by helping to direct resources into their most productive uses. Yet markets cannot function unless there are clear and transferable property rights and enforcement of contracts.

When property is held in common, institutions need to be developed, or even created, to encourage sustainable management. For instance, the allocation of irrigation water to competing users requires the establishment and operation of a competent authority. Responsible fisheries require a forum for cooperation among interested parties which has the power to enforce catch limits to prevent over-fishing.

In all such cases, the existence of an appropriate legal and institutional structure is clearly necessary for SARD.

4.5 Policy making that accounts for imperfect knowledge

4.5.1 Risk, risk aversion and downside risk

The very long-term nature of SARD makes decision making difficult. Errors in policy making or resource management may not come to light until far into the future. Moreover, bad decisions (including 'decisions' to do nothing) may lead to very bad outcomes (Ludwig, Hilborn, and Walters 1993). Many resource management choices are characterized by irreversibilities. For instance, an inappropriate land management strategy in an upland area may lead to serious soil erosion and land-slips. Once the soil is lost, it may be impossible to restore the land at any reasonable cost. Similarly, once a plant or animal species becomes extinct, it is lost for ever, no matter how potentially valuable it might have been.

Such a situation might seem to signal a need for very risk-averse policy making in resource management. Yet this is not necessarily the case. As Little and Mirrlees (1974) argue, in the context of project appraisal, the rejection of options just because they are risky could be a mistake. In so far as the consequences of policy decisions are spread widely and fairly evenly among the population of a country, many policy risks are not important. In many cases, though not in all, it is best to base the choice of policy, at least in principle, on what maximizes some measure of expected social value. 'Expected', in this context, means probability-weighted average computed across all possible outcomes, including any very bad outcomes that could occur.

There is an important distinction here between the notion of an expected value averaged across all possibilities, and a measure of social welfare calculated by assuming that 'average', 'best-guess' or 'typical' consequences follow from any policy decision. The former gives some weight to the risk of very bad outcomes which, as discussed, may be considerable when planning for SARD. By contrast, planning on the basis of 'normal' or 'most likely' outcomes means that the risk of such negative consequences is ignored.

The notion of planning to improve expected social welfare does require some judgements about the chances of occurrence of various kinds of consequences. The trouble is that the required probabilities are seldom known. Indeed, because of the long-term nature of planning for SARD, some actual outcomes may not even be conceivable at the time a decision has to be made. Planning in such an uncertain environment seems to call for extra caution. In addition, Little and Mirrlees note some special cases where risk aversion in policy making and planning is appropriate. A case in point is when the consequences of the decision are closely related to the overall performance of the economy. This may happen either because those consequences are large relative to the size of the economy, and/or they are strongly positively correlated with national income. For a small economy that is heavily dependent on income from fishing, such as some of the small Pacific island nations, for example, policy risks in the management of the fishery resources will obviously be more important than in larger, more diversified economies.

There may also be a case for risk aversion when a policy has the potential to have a sub-stantial effect on the welfare of certain groups, especially when those groups may be particularly vulnerable. Clearly, such a situation may prevail quite often in policy making relating to SARD. If policy fails to secure the sustainability of some agricultural system or rural economy, the local inhabitants may be seriously disadvantaged, even impoverished. In such situations, therefore, there is again a case for a more precautionary approach than basing decisions on expected benefits and costs alone.

Finally, special consideration needs to be given to cases where there are serious irreversibilities. In these situations, basing policy choice on expected values may seriously under-estimate the costs of future opportunities forgone when those opportunities cannot be quantified, and perhaps not even identified, at the time when a decision is taken. The need for a precautionary approach to policy making in this case is discussed in the next subsection.

4.5.2 Precautionary principles and adaptive planning

When any policy action or inaction carries with it the possibility of serious and perhaps irreversible resource or environmental degradation, with an associated risk of significant negative impacts on future welfare, actions to avoid or reduce that risk should be carefully considered. Such actions may include:

- postponing a decision to change the allocation or use of resources until more information is obtained about the possibility of serious negative results from the change;

- in a situation where continuation or projection of present practices creates the threat of serious negative consequences, relatively strict resource and environmental protection measures, such as safe minimum standards, may be imposed, at least until more is known.

The first rule might be called the 'look before you leap' principle, and the second might be called the 'better safe than sorry' principle.

In advocating these precautionary principles, it is not being suggested that policy makers should adopt a very risk-averse stance. Many risks to SARD, at least at the level of individual agro-ecosystems, are so small as to be insignificant in the context of policy making and planning at the national level. (Of course, the same is not true for threats to sustainability that may affect a wide range of systems, such as global warming.) Precaution is, however, called for when the impact on expected costs

and benefits of negative outcomes may be serious, in relation to the 'special cases' set out by Little or Mirrlees. It may also be appropriate when too little information is to hand to know whether the risks could be serious.

Operationally, the principles imply adding to the list of options usually considered by policy makers. Thus, the 'look before you leap' principle implies that, for example, instead of a choice between building a dam to generate electricity or not building it, the option is also explicitly considered of delaying any start to construction for a year or two. In that time, more data can be gathered on environmental impacts, and scope for energy saving that might eliminate the need to increase generating capacity can be explored. The 'better safe than sorry' principle means that, if there is uncertainty about the safety of a pesticide that is in common use, its use might be restricted until more conclusive evidence is available. Note that, in both cases, some actual or opportunity costs are incurred by acting cautiously, and these costs must be balanced against the achieved reduction in the chance of serious negative consequences.[1]

It might be argued that the very familiarity of the aphorisms used to characterize these precautionary principles shows that they are merely common sense, which indeed they are. Common sense, however, is not always applied in policy making, as many past policy mistakes reveal. Adding a requirement for caution to the agenda established for policy analysis may reduce the frequency of such mistakes.

4.6 The need for a systems approach to policy making

Any strategy to address SARD requires a comprehensive perspective that accounts for the interrelationships among the technical, environmental, social, economic, and political aspects of development. It must also provide for explicit recognition of, and understanding of, the social, economic and ecological diversity and uniqueness of each country, region and district.

A systems approach, involving multidisciplinary teams, offers the best possibility of

---

[1] For a formal treatment of the value of new information in dealing with irreversibility in resource management, see, e.g., Fisher and Hanemann (1990).

attaining a broad perspective encompassing these various dimensions. Moreover, there is a need for the strategy, and the choice of specific policy instruments, to be evolutionary. That means that decisions are adjusted or changed as more is learned about the nature of the problems to be addressed and about which solutions work and which do not.

The merits of the systems approach also extend to micro-level analysis for SARD. Selecting appropriate interventions for agriculture often requires understanding the responses of farm households to changes in the external environment. This understanding needs to be built on an appreciation of farm-household decision making. Again, therefore, a systems approach (FAO 1989) provides a relevant conceptual basis for micro-level analysis for SARD. For example, technological packages for SARD are mostly site specific. They depend on the farming system (including fisheries or forestry systems), size of units, resource stocks and quality, availability of inputs, access to markets, and socio-economic factors. The development and testing of improved technologies to promote SARD need to be tailored to identified main farming systems and validated on farms (or other units) representative of those systems.

5. STRATEGIC OPTIONS FOR SARD

5.1 General

A strategy is the way the policy campaign is managed. According to the Outline for Guidelines for FAO Policy Assistance in the Agricultural Sector, the following steps should be kept in mind when outlining such a strategy (FAO, n.d.):

- Consistency with national objectives, including the identification and, where possible, the resolution of conflicting objectives.

- Identifying overall goals of strategies, consistent with national goals, but narrowed to be operationally meaningful.

- Identifying components of the strategy, meaning the bundle of measures that must be taken to attain the stated objectives.

- Identifying areas of intervention, e.g., working out what is the proper role for government in promoting SARD.

- Identifying investment strategies, which follows from the previous point; many forms of intervention will require capital injection into, say, infrastructure, and it will be necessary to plan the nature, scale and timing of the required investments.

- Identifying policy interventions or reforms consistent with the overall strategy. This means correcting policies that are incompatible with the goals of SARD, and deciding what positive things government (or its agencies) can and should do to attain the objectives set.

Throughout the process of outlining an overall strategy, major concern should be given to *feasibility*. The strategy chosen must be technically and politically feasible, and within the capacity of the administration. Similarly, the investment strategy must be within the financial resources available. All this means that strategy development must be an interactive process in which both policy makers and stake holders are consulted. As already discussed, it will also need to be adaptive, adjusted as circumstances change and as understanding grows about SARD and the systems on which it must be based.

In selecting broad strategies for SARD, a number of important options need to be considered, as discussed below. Some of these options are real, in the sense that it is important for policy makers to make a choice of the right balance or compromise to match their particular circumstances. In other cases, the choice is more clear cut.

5.2 Growth or no growth

SARD is sometimes erroneously equated with zero growth in output. It is argued by those holding this view that agricultural intensification, requiring large quantities of off-farm inputs, is not sustainable. It is said that this form of production will deplete reserves of non-renewable resources, such as phosphate for fertilizer, and will lead to levels of pollution from chemicals used in production that cannot be absorbed into the environment without serious consequences. And, of course, there is more than a grain of truth in these views. On the other hand, as discussed earlier, the problem is that SARD must meet the challenge of the increased future demand for food and other agricultural products, and must provide sustainable livelihoods for increased numbers of rural people. There is no alternative: the process of intensifying agricultural production must be continued, at least in the main. Expansion of production is unavoidable and, indeed, is necessary to meet the criteria for SARD.

At issue, therefore, is not whether there should be growth or no growth. Rather it is a matter of how the required growth can be achieved without degrading either the agricultural resources or the environment (de Haen and Saigal 1992). That is, the challenge for policy making for SARD is to find ways to promote changes that lead to sustainable agricultural growth.

Of course, there are agro-ecosystems that are so fragile that further sustainable intensification is not possible, at least with present knowledge and at reasonable cost. But equally, many more agricultural systems could be improved to yield more output while, at the same time, they are made more sustainable. So, placement of mineral fertilizer rather than broadcasting may lead to both higher yields and reduced pollution of ground water. Integrated pest management can be at least as effective as spraying with insecticides, as well as cheaper. Responsible fishery management yields a higher and sustainable catch, with less effort, than is possible with unrestricted access and over-fishing.

5.3 Intervention, incentives and control options

In an ideal world where markets work perfectly, scarce resources would be allocated among competing ends in ways that maximize the benefits for all (present and future) members of society. Even in the real world, as some resources become scarce, their prices rise, signalling to resource owners the need to use less of them, perhaps by substituting other inputs. Thus, as land for farming grows more scarce, its price rises and farmers find it profitable to apply more labour and capital to given areas. The changing costs of production associated with changes in resource availability are reflected in the quantities of goods and services producers are prepared to offer for sale at particular prices. As a result, consumption patterns adjust accordingly. But in addition, consumers' preferences for particular bundles of goods and services are communicated to producers of those items through the prices the consumers are prepared to pay. Hence, the resulting prices and quantities traded reflect the interaction of consumers'

preferences with production costs, including the costs of using scarce resources. Finally, investors and resource owners try to balance profits today against the possibility of more profit in the future should increasing resource scarcity drive up prices. Such inter-temporal tradeoffs are governed by the operation of the capital market that signals to investors when to invest and when to refrain from investing.

In theory, therefore, a perfect market should lead to the optimal use of resources both now and in the future, thus bringing about sustainable development. So, if markets really did work in this idealized way, there would be no need for policy intervention to attain SARD.

Unfortunately, however, there are several ways in which markets can fail. Important causes include:

- intended and unintended consequences of government interventions, such as taxes and subsidies;

- attenuated property rights associated with some common property and most open access resources;

- externalities;

- imperfect and asymmetric information;

- monopolistic competition;

- imperfect capital markets; and

- the inadequate capacity of market forces to accommodate value judgements about current and inter-generation equity.

Each is discussed in turn in the next sub-section.

5.3.1 Types of market failure

Attenuated property rights. Attenuated property rights—implying inability to enforce ownership rights—are common in agriculture (including forestry and fisheries). Land, trees or fish stocks may be held in communal ownership, with individual households having use rights only. Alternatively, there may be open access to some resources, with everyone having rights to use the resources. In either case, resource users have diminished incentives to preserve or develop the capacity of the resources. The attenuated property rights may mean that individual investors cannot capture for themselves all the benefits of investment in resource improvement. Or it may mean that they do not bear the full costs of allowing the resources to be degraded or depleted.

The well-known 'tragedy of the commons' occurs mainly with open access resources (and also with some common property resources). With no regulation in place, each individual using the resource earns the average revenue, and not the marginal revenue, attributable to his or her use. Open access to fishery resources, for instance, commonly leads to their over-exploitation. Similar problems can occur with common property resources; here, institutional control over these resources often breaks down under pressure from population growth and commercialization. Loss of control by institutions responsible for governing the use of common property resources can lead to an increased number of disputes, and failure to enforce sanctions on those who violate resource-use restrictions.

Externalities. Externalities occur when resource users create costs that others have to bear or, less commonly, where they create benefits for which others do not have to pay. For example, upland farmers may so manage their land as to cause damage to water courses affecting downstream water users. Usually, there is no market mechanism in place to force the upstream people to pay the costs of the damage they inflict on those downstream, nor any way the downstream people can offer incentives to those above them to mend their ways.

Imperfect information. Markets also fail when market participants have imperfect information and understanding—probably the norm in many agricultural systems. Environmental changes may be slow, at least at first, and may not be detected without close monitoring. Consequently, local people may not be aware of what is happening to their resources or, more plausibly, may lack the knowledge to do anything about it. Thus, markets for land or other resources, may imperfectly reflect future productive capacity.

The market failure is more likely to be severe and distorting when there is asymmetry of information between buyer and seller. Logging contracts between overseas companies and local individuals, communities and even governments are a common example. The locals may have little or no knowledge of the true value of their timber resources—a state of ignorance that the loggers work to preserve by such measures as 'transfer pricing'. (The latter is a method whereby the exporting firm

covertly 'sells' the timber to itself for less than it is worth.) In consequence, the contracts are too often written on terms unfavourable to the timber owners, leaving too little surplus to pay for replanting and other necessary post-logging conservation measures.

Monopolistic competition. The power of monopolies to manipulate market situations for profit is well known. Accusing fingers are commonly pointed at the large, international corporations, which are certainly no paragons of virtue and may well use their market power in ways that cause rural resources to be depleted. But governments also may limit competition in agricultural markets, for example by granting exclusive trading rights to statutory marketing authorities. If such marketing institutions grow inefficient or corrupt, the markets they control no longer convey the right price signals to farmers, leading to inefficient resource use, and so detracting from sustainable development.

Imperfect and distorted rural capital markets. Imperfect and distorted rural capital markets are a hallmark of rural production in many less developed countries. Such markets emit the wrong signals to agricultural producers by pushing up the real cost of capital. The higher the interest rate, the less attractive it is for producers to invest in resource conservation, and the more attractive it is for them to run down resource stocks. As Lipton (1989, p. 1) noted:

> When the rate of interest is very high—and poor people in the rural Third World commonly pay 25-50 percent in real terms—investment in long-term conservation of private property is not appealing: 'soil mining' is.

Equity and the market. Central to the sustainability debate are concerns about the capacity of market forces to accommodate value judgements about current and, particularly, inter-generation equity. The power of the poor to command goods, services and resources in competitive markets is clearly limited. The untrammelled operation of competitive markets will not lead to the attainment of policy goals relating to poverty reduction and improved equity. This form of market failure is all the more important in the attainment of the goal of reducing poverty and deprivation of future generations—the unborn are powerless in the competitive markets of the present.

To sum up, these various forms of market failure are perhaps the more important part of the reason why policy intervention for SARD is an issue deserving careful consideration. Yet the existence of market failure does not automatically justify policy intervention, as discussed next.

5.3.2 Policy failure

Governments and their policy advisers often have inflated expectations about their capacity to change the way people behave and use (or abuse) resources. Policy interventions are often ill-conceived or inadequately implemented. As a result, their impacts are too often ineffective, sometimes even contrary to intentions.

Some common failures in policy include:

- Failure to recognize the impacts of macro-level policies on the micro level.

- Over-reliance on regulation to the neglect of solutions based on ameliorating market failure.

- Focus on larger-scale, commercial producers to the relative neglect of the resource poor.

- 'Compartmentalization' of policies, including agricultural and environmental policies.

- Over-emphasis on excessively optimistic resource use planning relative to knowledge development.

There is a widespread tendency for policy makers to turn to command and control regulations when it appears that market failure is leading to less preferred outcomes for SARD. Yet there are good reasons why such regulations seldom work, especially (but not solely) in less developed countries (Panayotou 1994b):

- Draconian penalties for breaches of environmental regulations are seldom enforced and are often unenforceable.

- It is usually impossible to monitor the myriad small-scale agricultural producers who control most rural resources. And, even if technically possible, the costs of monitoring and enforcement would make the exercise uneconomic.

- There is usually a mismatch between the implied requirements for monitoring and enforcement of regulations and the budget-

ary, personnel and administrative capacities of less developed countries.

- Actual penalties imposed on violators of regulations who are prosecuted are often so low that, given the low probability of apprehension, they do not serve as a significant deterrent.

- Governments often lack the capacity and willingness to use education and persuasion to change social attitudes in favour of compliance. Yet such attitudinal change may be needed so that those who break the rules are stigmatized in their own communities.

- Command and control regulations elicit rent-seeking behaviours, such as bribing of officials, that not only undermine the intent of the regulations themselves, but may lead to corruption of the whole process of administration.

5.3.3 Picking a strategy to deal with market failure

The foregoing discussion of market and policy failures signals a need for SARD policy makers to make some broad strategic judgements. They need to consider the merits of more versus less intervention (zero intervention across the board will never be a realistic option), and the relative emphasis placed on market-based incentives versus command and control measures. While such broad strategic choices will be made in part on the basis of political views, some comments bearing on the choices can be offered.

Perhaps the first rule might be 'If it ain't broke, don't fix it'. In other words, if present arrangements, whether market-based or interventionist, appear to be operating reasonably consistently with SARD objectives, there is no need to change unless something truly better can be put in place.

When the existing arrangements are thought to be unsatisfactory, the next question to ask is whether it is feasible to design a policy intervention that will improve the situation. The information required to identify what is going wrong and why, and to devise a policy to correct the situation may not be obtainable at reasonable cost. Often, considerable uncertainty surrounds predictions about how the various agents will respond to a given policy initiative, so that its feasibility and efficacy will be doubtful. It is for this reason that simple, reliable interventions may be better than more ambitious but untested ones.

If and when it is thought that a feasible policy intervention has been identified, the next question is whether the benefits from the policy will justify the costs of implementation. These costs include the direct administrative costs faced by the government or the implementing agency, and also include 'hidden' costs of compliance by those affected by the legislation. Thus, a decision to ban a particular insecticide may prevent suspected adverse effects on the environment and/or on human health. On the other hand, it may also add to the costs of agricultural production through reduced yields or through the need for more expensive pest control measures.

Clearly, only those interventions that are needed, that are technically and administratively feasible, and for which benefits are expected to outweigh costs, should be implemented. If application of this rule leads to less rather than more policy intervention, that may be no bad thing. With less to do, governments and their agencies may be able to manage the more limited range of tasks more successfully.

A good starting point for policy might be to scrap, or redesign, existing policy interventions that are not consistent with SARD. Thus, it is not uncommon to find government-created distortions or impediments that are inhibiting the functioning of competitive markets, the removal of which would help assure more sensible resource use. Some specific policy instruments to these ends are discussed in section 6 below.

Given the prevalence of various forms of market failure as causes of unsustainable agricultural practices, and the perceived difficulties in making regulations work, many economists have argued that policy making for SARD is often best approached by fixing the market failure. That may mean finding ways to make failing markets work better, or it may mean finding a way to 'get prices right'. There are a number of ways in which these notions can be implemented.

For example, where a lack of enforceable property rights is leading to over-exploitation of a resource, as happens with open access and some common property resources, there may be a case for privatizing such resources. The privatized resources might then be allocated to the existing users. However, while privatization may lead to better management of

the physical resources whose ownership is transferred, privatization of common and open access property may have a negative impact on equity (Baumol and Oates 1988, ch. 15). Not surprisingly, therefore, privatization decisions are often highly politicized.

Alternatively, some market failures may be handled by setting a tax or subsidy to 'correct' distorted prices. One example of this approach is the so-called 'polluter pays' principle, whereby those causing negative externalities through pollution are taxed in proportion to the damage they cause. Similarly, positive externalities from agriculture or forestry, such as contributions to such public goods as landscape, biodiversity and wildlife habitat, might be rewarded by subsidies. These might be funded partly by charges on beneficiaries such as recreational users of the land. Some examples of such policies, and their advantages and limitations are also discussed in section 6 below.

There are, however, limits to the applicability of the notion of adjusting prices for market failure; in practice, it is not possible to 'create' a market for every good (or 'bad') to mirror the 'right' price for SARD. However, if the 'right' price can be estimated by some other means, it may still be possible for policy makers to evaluate the impacts of alternative interventions using the appropriate social values. Most obviously, in project appraisal, prices used for economic or social analysis should reflect, so far as possible, the true costs and benefits to present and future generations of the alternatives considered. Similarly, there has been much concern that national income calculations do not properly measure the changes in levels of resource stocks and environmental quality. Approaches to full resource accounting have been developed to try to correct these deficiencies.

The valuation of resources and the environment presents some interesting challenges for economists. There is a large and growing literature on this topic that is too voluminous to review here (see, for example, Markandya 1995; FAO 1994a; Pétry 1995a, b). Policy advisers concerned with SARD will want to be familiar with the latest thinking on the topic.

Finally, despite the weakness of command and control approaches to deal with market failure, there are situations when regulations are the best, even the only, way to deal with market failure that threatens SARD. For example, absolute bans may need to be imposed on the importation of agricultural items such as plants and animals that may be the source of serious pests or diseases that could seriously affect the sustainability of local agriculture.

5.4 'Top down' versus 'bottom up' approaches

It is wrong to suppose that SARD is wholly or even mainly a matter for policy makers. In most countries, the important decisions about agricultural resource use are made by the rural people, particularly, but not exclusively, the resource managers. It is the farmers, forest people and fisherfolk who are responsible for most of the resource utilization decisions that affect SARD. These rural people are, of course, responding to the demands for agricultural products expressed through the prices that consumers are willing to pay for those products. The attainment of SARD requires the creation of conditions for both producers and consumers to change their ways.

What governments can try to do is to create the conditions and establish or improve local organizations to encourage people to change to more sustainable patterns of production and consumption. At the same time, by encouraging the participation of people in the process of SARD, government policy makers can learn from the people about their needs and circumstances, and about their likely responses to policy initiatives. This is another way of saying that a strategy for SARD needs to be both a 'top-down' and a 'bottom-up' process. Policy makers need to find culturally acceptable and effective ways of involving people in their decisions and of motivating people to search for and adopt their own solutions to problems of unsustainability. Some specific policy options are considered below.

6. POLICY OPTIONS AND INSTRUMENTS

Policies that affect sustainability are of five types (FAO n.d., 48-9):

- General economic and social policies intended to influence overall economic growth, trade, price levels, employment, investment and population, attained chiefly by utilizing monetary and fiscal instruments.

- Policies relating to agricultural and rural development. Policies of this type are usually intended to influence such factors as the agricultural resource base, agricultural production, consumption of agricultural products, agricultural price levels and variability, rural incomes and the quality of food. They are usually implemented via instruments such as taxes and subsidies, direct government production and provision of services, and direct control through regulation.

- Policies relating to markets, including the establishment of market institutions and rules, and circumscription of property rights.

- Policies aimed at establishing a democratic and participatory process designed to involve all interested groups in decision making and implementing SARD.

- Policies designed specifically to influence natural resource use and protect the environment. These policies utilize: (i) command and control (effected, for example, by prohibiting or limiting certain resource uses or establishing limits on emissions, with penalties for non-compliance); (ii) economic incentives such as taxes and subsidies; and (iii) persuasive measures such as education and advertising.

The first four categories above are not primarily intended to achieve SARD, and are adopted to accomplish other goals. Yet all are essential for SARD. Therefore, the challenge in policy making is to integrate sustainability and environmental considerations into mainstream policy making both within the agricultural sector and generally.

In the following sub-sections, these five main categories are discussed in more detail.

6.1 General economic and social policies

6.1.1 Fiscal and monetary policies

Fiscal policies include government expenditure, taxes and subsidies. Direct impacts on SARD arise from expenditure on such things as agricultural research and extension, and public works in rural areas. Taxes, on the other hand, may be targeted to help regulate resource use, such as resource rent taxes or taxes on polluters. Some specific examples of targeted expenditures and taxes are discussed under later sub-headings.

Many governments, especially in industrial-ized countries, raise the majority of their taxes from progressive income taxes. If net incomes of workers are to rise with general economic growth, labour costs to employers must rise more steeply to cover the extra tax. Faced with a higher wages bill, firms will find it profitable to substitute other, less heavily taxed natural resources for labour. Clearly, such a taxation system will lead to a too high utilization of natural resources, renewable or otherwise, relative to labour. Biasing the taxation system more strongly towards resource taxes should help promote more sustainable development. In the many developing countries where income taxes are not that pervasive, the issue will be less the tradeoff between income and natural resource taxation than between the latter and other forms of taxation.

More generally, in the short to medium term, prudent fiscal management is important for SARD. The reason is that, without it, economic instability may be created that is likely to discourage private investment, and frequent budget crises may disrupt the provision of important public services. In the longer term, a too large public sector, with high rates of taxation and high government spending, may 'crowd out' development in private sector activities, of which agriculture is usually one. Also in the longer term, the way the government allocates expenditure among sectors and activities, and the way taxes are levied differentially on various sources of income, activities and assets, influences both the sectoral make-up of GDP and the distribution of income. Thus, governments that allocate resources mainly to the urban areas, and tax agriculture, directly or indirectly, will discourage SARD, whether wittingly or not (Markandya and Richardson 1994).

Monetary policies can affect inflation, interest rates, and credit supply. The effect of inflation on SARD is likely to be complex and not easily predicted. However, measures to limit inflation, such as price controls placed on basic items like food, will obviously damage the agricultural sector and so are likely to dis-courage SARD. Moreover, inflation leads to economic distortions that, in the main, are also likely to be inconsistent with the attainment of SARD.

The general view about the relationship between interest rates and SARD is that low rates are preferable to encourage resource owners to make investments in resource con-servation measures that yield benefits only in the longer term. However, if low interest rates are achieved by artificially holding down rates, the result may be 'financial repression', meaning that the development of the financial sector will be inhibited. This will discourage formal savings and so will limit the funds that financial institutions can lend as credit. Rural people are often 'at the end of the line' for the delivery of credit by formal financial institutions. As a result, the rate of rural invest-ment tends to be slowed, with negative impli-cations for SARD.

6.1.2 Trade and exchange rate polices

This group of policies includes taxes on imports and exports and controls on trade. Such policies tend to reduce the domestic producer prices for export products and to push up the domestic prices on imported items. These impacts will affect commodities and sectors differentially, and it is impossible to say for sure what the effect on SARD will be. However, in so far as export taxes are imposed on agricultural products, rural incomes will suffer. The damage may happen directly and also through the effects on prices of domestically produced food as producers switch away from production for export. Because SARD is partly about sustainable rural livelihoods, interventions that reduce rural incomes also damage sustainability. Import duties on agricultural inputs, or restrictions on their importation, are likely to have a similar effect, except in the case of inputs such as agricultural chemicals that cause serious environmental and human health risks.

In a global setting, a more open foreign trade framework, with fewer taxes and restrictions on imports and exports, is usually held to be good for sustainable development. The case for free trade is that it will allow production to take place in accord with the economic principle of comparative advantage. That in turn would mean that less resources and other inputs would be needed in aggregate to attain a given level of production, which should be good for everyone and for the environment. However, as discussed above in sub-section 4.1, free trade, while generally having the expected benefits on economic growth and efficiency, may have negative effects on equity, at least at first. The environment too may suffer if, for example, free trade makes it possible for rich countries to export some of their pollution problems to poor countries.

So far as exchange rate policy is concerned, some countries, concerned about the impact of negative external balances on their international purchasing power, have sought to maintain a high official exchange rate by limiting imports. Experience and economic logic both suggest that such policies are likely to be unsuccessful in the longer run. However, while they are in place, they turn the domestic terms of trade against those sectors of the economy that produce tradeable goods, including agriculture. And the distortions induced can be massive, far outweighing any subsidies that may be offered to farmers such as bounties on fertilizers. Clearly, such distortions can be devastating for SARD.

6.1.3 Labour and employment policies

Wage policies normally cover only that part of the labour force that is unionized and in the formal sector. In many less developed countries, the proportion of the labour force in such employment is small, so that wage policies have limited impact on either overall wage levels or SARD. (The same may not be true for the industrialized countries.) However, wage policies may affect the allocation of labour between sectors. Moreover, if large differentials between urban and rural wages result, they may lead to an unwanted acceleration of rural-urban migration, with concomitant problems of urban congestion and pollution.

Also of concern is the issue of rural unemployment, whether overt or hidden, existing in some countries. In countries where social welfare programs are not well developed, under-employment or unemployment among rural or urban workers is a serious threat to sustainable livelihoods. While solutions may be sought in the medium to longer run by encouraging economic growth and the development of labour-intensive industries, in the short run, relief measures may be needed. 'Food for work' schemes may provide an opportunity to combine the relief of poverty with labour-intensive public works. Such works can be used to conserve or enhance the resource base, for example through afforestation programs or public works to control soil erosion.

Aspects of employment policy relating to human resource development are discussed later.

6.1.4 Investment and foreign aid

Given that the goal of sustainability policy is to transfer the equivalent of the current resource base to the next generation, investment policy is obviously crucial. In part, this is a matter of creating an environment that is 'friendly' towards private investors. Too many barriers for the entry of foreign investors will deter the international flow of capital, as will too strict rules on the repatriation of profits. Similarly, both economic instability and social unrest will deter all investors, foreign or local.

On the other hand, investments can be devastating for SARD if investors do not have to pay for negative externalities they cause (e.g. dams taking water from downstream users, wells lowering the water table for all). Clearly, measures need to be in place either to prevent such negative externalities, or to make investors pay for them.

The importance of avoiding interventions that cause financial repression and so deny would-be investors, especially those in rural areas, access to credit markets has already been discussed.

Because public funds are always limited, direct government investment should be allocated to areas where market failure leads to under-investment. Such areas include investments in human capital (i.e. in people). Building human capital for sustainability is likely to require substantial government investments in health services, education and training. Similarly, public investments may be needed in open access resources such as many fisheries and some forests, as well as in state-owned resources such as rural infrastructure. (See Bromley and Cernea 1989 for a discussion of the different types of resource property regimes.) For SARD, an appropriate proportion of such improvements needs to be directed to rural areas where these services are typically very inferior to those in the cities.

Common property and open access resources important for SARD may include fisheries, forests, and water supplies. Some state property, such as public roads, also has open access characteristics. Because individual private investors are seldom able to capture for themselves all the benefits from investments in open access resources, there is often a need for intervention. Governments may need to step in either to undertake those investments that are socially profitable but privately unprofitable, or to create institutional arrangements whereby externalities are internalized to enable private investors to earn an appropriate return on their capital. Examples of the latter type of policy intervention include the privatization of water supply agencies and the letting of contracts for the construction and operation of toll roads.

Government investment decisions need to be based, so far as possible, on a careful assessment of the social benefit of each investment. The methods for undertaking such project appraisals are well known, and have been extended in recent years to try to take proper account of environmental impacts (Edwards-Jones 1996; World Bank 1991). However, as discussed earlier, performance measures based on discounted cash flow may need to be used as part of a multi-criteria analysis that also includes other aspects such as inter-generation equity (see section 8.5.3).

In less developed countries, where government funds are typically very limited, foreign aid, including concessional loans, may be used to help make good any shortfall in the level of public investment needed for SARD.

6.1.5 Population policies

The threat to SARD posed by population growth was indicated in section 3.1 above. It is clear that, at least in the long run, population growth must be slowed and perhaps reversed if sustainability is to be attained. Even in the shorter run, population growth presents challenges for many less developed countries. Substantial investments are needed in order to expand both food production and the provision of services such as education and health in line with the increasing numbers. Sustainable development is therefore made all the more difficult in such cases.

The trouble is that some of the more reliable birth control measures are not acceptable in all societies. This tends to make the whole subject of population policy a difficult one for policy makers. Yet there are other means of approaching the topic than by a head-on confrontation with sometimes strongly held ethical values.

Policies that improve the education of girls and the status of women, including the employment prospects for younger women, will usually lead to a reduction in fertility. Moreover, when the status of women rises, they are more likely to take more control of their own fertility, no longer relying on the decisions of their male partners, local health workers or religious leaders.

More generally, there is evidence that birth rates tend to fall as child mortality is reduced, urbanization spreads and living standards rise. The policy interventions to promote such outcomes are seldom contentious, and will slow population growth.

That said, it is held by many to be a desirable policy to make sure that no woman is denied access to the means to limit the frequency with which she conceives, regardless of religious or other taboos that dictate what methods are available to her to attain that end.

Population policy also extends to measures to affect the geographical distribution of people. Governments can influence the rate of migration to urban areas, chiefly by manipulating, wittingly or otherwise, rural-urban real income differentials. Policies to concentrate public investment and service provision in the cities will accelerate rural-urban migration, and vice-versa. Similarly, governments can adopt policy measures to try to influence the distribution of the rural population, for example through land settlement schemes such as the Indonesian transmigration programs.

6.1.6 Incomes and equity policies

The importance of taxation in affecting incentives to use resources in a more or less sustainable way has been discussed in 6.1.1 above. A second consideration in setting taxes, however, relates to the distribution of income and wealth. Taxing the rich and stripping them of some of their assets to meet the needs of the poor of this generation, or to invest for the benefit of poor people in the future, can obviously be one way of attaining the goal of sustainable development. Unfortunately, it is seldom politically feasible on any substantial scale! However, where wealth or income comes from economic rents that result from public investments, such as increased value or productivity of land due to public infrastructure construction, it may be more acceptable to tax away at least a part of those gains.

Policies can also be adopted that specifically target the poor, such as famine relief programs, food for work schemes, or provision of free or subsidized services. Most industrialized countries have extensive social welfare programs in place to help the disadvantaged, such as the sick, the old or the unemployed. Because such programs are very expensive, in most less developed countries they are of much more limited scope, or are even totally absent.

The argument for policies designed to achieve a more equitable distribution of income is not that the poor cause more resource degradation than the rich, and so must be helped out of their poverty. Indeed, the contrary is often the case, given the much higher levels of consumption of items such as fossil fuels in rich countries. Rather the aim is to make possible a reasonable standard of living for today's poor. Then they can have the means to provide a more sustainable future for their children, for example by providing them with good nutrition, sound health care and a good education.

6.2 Policies relating to agricultural and rural development

While the general economic and social policies outlined above may be very important for SARD, they mostly fall outside the area of responsibility of agricultural professionals. The policy areas discussed next, on the other hand, relate specifically to the rural sector and will, in many cases, lie within the competence of ministries of agriculture.

6.2.1 Rural infrastructure

Rural infrastructure improvements contribute to SARD by improving the availability of services or other facilities that enhance the productivity of private rural capital. A tarmac road that lowers travel time and reduces vehicle running costs, also lowers the costs of marketing agricultural produce and reduces the delivered cost of farm and household requisites. Similarly, an irrigation system raises the productivity of farmland while a telephone system lowers transaction costs in agricultural marketing and improves efficiency by giving producers better access to price information.

While public investments in rural infrastructure are important for SARD, so is the provision of systems and funding for the maintenance of existing infrastructure. Without the latter, investments in infrastructure may be wasted or may yield lower benefits than they should. One of the most significant impacts on sustainable development of the crisis in Africa in the 1980s was the serious deterioration of all types of infrastructure because of lack of maintenance.

Because provision of infrastructure is usually expensive and because opportunities are many and funds restricted, major policy questions arise as to which infrastructure projects are to be given priority. In such a situation, proper investment appraisal is important. Such appraisals should obviously account for environmental impacts of the proposed project. In planning a new road, for instance, proper attention to alignment and to such design features as the control of storm-water run-off may add little to costs but may reduce the environmental harm considerably.

Given that the planning is done well, infrastructure improvements can promote SARD:

- by improving the terms of trade of producers in previously isolated areas;

- by permitting the new or more intensive use of previously too isolated land or other resources, so reducing pressure on other areas;

- by permitting previously isolated rural communities to benefit from better access to services such as health and education, and agricultural extension.

6.2.2 Building human capital for the rural sector

Human capital improvement is the aspect of SARD that can too often be neglected in the thinking of agriculturalists. Yet, as noted earlier, improvements in human capital have been the source of most of the gains in the productivity of agricultural land and labour in the past. Given that the land frontier has been reached in most countries, and that areas available for farming and forestry are likely to decline, policies to enhance rural human capital need to be given high priority.

The provision of universal education and health services in rural areas are two main ways of improving the agricultural human resource base. Usually, neither of these will be the responsibility of agricultural planners. However, the training of agriculturalists and the dissemination of agricultural knowledge to farmers and others is usually a matter in which agricultural ministries have some responsibility.

Policy interventions here, therefore, relate to the provision and operation of training facilities, and programs to extend that training into rural villages. The scope, organization and orientation of agricultural extension also need to be considered. The existing extension service may not be up to the task of promoting SARD. The skills of the personnel may need to be upgraded to provide them with better understanding of sustainable farming methods. Given the increasing recognition of the importance of people's participation in attaining SARD, those extension services that have been used to a 'top-down' approach may need to undergo dramatic reform.

The role of the mass media in promoting SARD should also be considered in reviewing policy. More sustainable farming practices may be promoted using media such as radio which, along with other promotional methods, may be a relatively cheap and effective way of changing attitudes and values about rural resource use.

6.2.3 Agricultural research and technology development

The contribution of improved technology to SARD has already been emphasized. Policy issues relate to how the development of such improved and sustainable technologies is to be encouraged. At issue in the public sector will be the level of funding for agricultural research and the allocation of those funds between research organizations, commodities, disciplines and projects. The merits of a farming systems approach versus a more traditional disciplinary division of activity needs to be examined. Whatever approach is taken, mechanisms for setting research priorities that are relevant and responsive to the needs and circumstances of rural producers need to be in place. The research done also needs to be reviewed regularly to ensure that funds are being used wisely and well, and that progress towards the intended goals is being attained.

Policy attention also needs to be given to the staffing of R&D efforts for SARD. That means determining the allocation of funding for agriculture faculties of universities and for agricultural colleges. It also raises the question of whether research training in various sub-disciplines is to be provided in-country or using overseas universities that may be better equipped for the task. In many less developed countries, the retention of trained research staff may be a problem, given the international mobility of good scientists. Measures such as bonds and incentive payments may need to be put in place to secure the proper staffing of research programs.

In addition to public sector R&D activity, policy initiatives to promote private sector efforts should also be considered. Instruments here include possible government support for private or corporate research organizations, perhaps by raising levies on agricultural sales. Attention also needs to be given to property rights legislation, such as patents laws and plant variety rights, in order to give innovators the opportunity to profit from their investments. Possible tax breaks for firms doing R&D may be another form of intervention that may generate social benefits and promote SARD.

The detailed consideration of such agricultural research policy and management issues as those broached above is outside the scope of these guidelines. The International Service for National Agricultural Research (ISNAR) is a centre within the CGIAR set up specifically to assist in such matters. ISNAR has published a number of relevant guides (e.g., Chissano 1994; Crosson and Anderson 1993).

6.2.4 Agricultural prices

Policies designed to raise the prices paid for agricultural produce may be implemented by restricting or taxing competing imports, by limiting production by quotas, or by direct subsidies. Producer prices may be driven down by restricting or taxing agricultural exports, by subsidizing imports, or by directly taxing sales.

In the past, at least, governments in less developed countries have typically sought to drive farm prices down to keep food prices low. Often, the aim was to benefit urban dwellers and to restrain the growth of urban wages. Governments in industrialized countries, on the other hand, have typically sought to push farm prices up to satisfy farm lobby groups. Recently, there has been increasing recognition that both types of distortion cause misallocation of resources and are therefore not conducive to SARD.

While any inefficiency in resource allocation can be damaging for SARD, there are some specific negative impacts of distorted agricultural prices. Low agricultural prices threaten SARD by discouraging the growth of farm production and by making it difficult for rural people to earn sustainable livelihoods. Investment in agricultural resources, such as land improvements, will be dampened under a low-price regime. High prices, on the other hand, lead to uneconomic use of inputs, some of which, such as chemicals, may be damaging to the environment. High prices may encourage the too intensive use of marginal lands. They may also prompt socially unprofitable and unsustainable but privately profitable investments, such as unsuitable clearing of forest and woodland, or draining of wetlands, for agricultural use.

6.2.5 Stabilization and risk in agriculture

There are two main areas of policy intervention in relation to risk in agriculture. The first is in hazard reduction. Major natural disasters, such as cyclones, severe floods, or fires, can have significant negative impacts on the resource base. Land may be irreversibly damaged by sudden erosion, land slips or inundation. Similarly, assets such as crops, trees, animals, and land-based improvements such as fences, terraces, irrigation works, roads and villages may be damaged or destroyed over large areas.

Policy makers need to give attention to justifiable measures to prevent or limit the impacts of such disasters. Instruments include the construction of protective works such as levee banks or fire breaks, and the establishment and maintenance of disaster warning systems.

The invasion of agricultural systems by serious pests, diseases or weeds is another potential hazard that can be minimized by proper quarantine measures and by having in place plans to control outbreaks before they spread.

Man-made disasters, such as wars and civil unrest, can produce damage at least on the same scale as many natural disasters. In these cases, both prevention and mitigation of consequences are obviously more difficult.

The second type of policy intervention is needed after a disaster, whether natural or man-made, has happened. For the damaged systems to recover with minimal threat to sustainability, governments (perhaps along with international agencies) may need to implement various forms of disaster relief. Relief measures adopted may include emergency food supply, provision of materials for replanting crops, supply of replacement breeding animals, and help with reconstruction.

6.2.6 Direct government involvement

Governments often engage directly (usually through government agencies set up for the purpose) in rural resource management, more often in forestry than in farming. They also often participate in agricultural input supply, provision of rural credit, and in marketing agricultural production.

In regard to all such participation, the policy question to be addressed is whether government agencies are as efficient and effective in the attainment of SARD as private enterprise would be. Governments are increasingly moving to privatize these types of functions in the belief that the answer to the question is in the negative. However, it is unwise to be too dogmatic on the issue. There may be cases where public ownership can be advantageous. For example, governments can typically get access to capital at lower rates than private businesses. Arguably, therefore, they are better placed than the private sector to make the long-term investments needed for sustainable production in, say, forestry. Thus, direct government participation in production and marketing in agriculture needs to be looked at on a case-by-case basis, and not judged in any doctrinaire fashion.

6.2.7 Sustainable rural livelihoods

Since SARD means sustainable rural livelihoods, now and in the future, policies are needed in many less developed countries (and in some industrialized ones) to deal with the persistent problem of rural poverty. Unfortunately, it is not easy to identify and implement appropriate remedial measures—if it were, the problem of poverty would have been solved long ago.

Sustainable reductions in poverty will surely require the integrated and effective implementation of a wide range of policy initiatives under all or most of the headings in this section of the guidelines. To illustrate what may be entailed, Box 7 contains one suggested list of the policy measures needed to assist the rural poor to attain sustainable livelihoods.

Box 7

SUSTAINABLE RURAL LIVELIHOODS FOR THE 21ST CENTURY: POLICY IMPLICATIONS

Adapted from Chamber and Conway 1992, pp. 31-33.

I. Enhancing capability:

The provision of enabling infrastructure and services including:

- education for livelihood-linked capability
- health, both preventive and curative to prevent permanent disability
- bigger and better baskets of choices for agriculture, and support for farmers' experiments
- transport, communications and information services (about rights, markets, prices, skills...)
- flexible credit for new small enterprises

II. Improving equity

Giving priority to the capabilities and access of the poorer, including minorities and women, via:

- redistribution of tangible assets, especially land, and land to the tiller
- secure rights to land, water, trees and other resources, and secure inheritance for children
- protection and management of common property resources and equitable rights of access for the poorer
- enhancing the intensity and productivity of resource use, and exploiting small-scale economic synergy
- rights and effective access to services, especially education, health and credit
- removing restrictions which impoverish and weaken the poor

III. Increasing social sustainability

Reducing vulnerability so that the poor do not become poorer by:

- establishing peace and equitable law and order
- disaster prevention
- counter-seasonal strategies to provide food, income and work for the poorer at bad times of the year
- prompt support in bad years, and high prices for tangible assets people sell in distress
- health services that are accessible and effective in bad seasons, including treatment for accidents
- conditions for lowering fertility.

6.2.8 Food and nutrition

The provision of an adequate diet and satisfaction of nutritional needs for all are inseparable from SARD. As discussed in subsection 3.2, population growth and rising incomes will cause substantial increases in the demand for food. Moreover, progressive reductions in the numbers of poor people who often go hungry are necessary for sustainable development, to reduce suffering and to provide better opportunities for them to live healthy, productive and fulfilled lives.

Food and nutrition policy goals include security of food supply, safe and good quality food and adequate and healthy diets for everyone. To a large degree, these goals are consistent with the broader objectives for SARD. Sustainable agricultural production will mean increased food supplies and increased income-earning opportunities, leading to reductions in poverty and malnutrition. In some countries, however, specific measures to improve the food security of the poor and malnourished may be needed. The appropriate policy interventions here are the same as those mentioned in relation to the previous topic of sustainable livelihoods.

A focus on food security also means correcting any imbalance in the proportion of agricultural research and extension efforts directed to food crops, rather than to export or industrial crops. In the past, there has been a tendency in some countries to neglect roots, tubers, plantains, traditional legumes, oilseeds, vegetables and fruits, which are commonly eaten but which less frequently enter domestic and international trade. Promoting the increased production of some of these crops may lead to better human diets and more sustainable production systems. The reason is that such crops can provide alternatives to the intensive production of the main staples or export crops. More attention may also need to be given to post-harvest systems for all food crops that are affected by significant losses and quality degradation.

Measures to improve the quality and safety of food have direct beneficial effects on health and nutritional status. They include establishing and enforcing food laws, support for consumer organizations, and education to improve food quality, safety and preparation. Measures are needed to prevent the contamination of food with agro-chemicals, or by pests including micro-organisms, as are procedures to detect such contamination in locally produced or imported foods.

Widespread nutrition and diet education, delivered through formal schooling and the mass media, can promote good eating habits and healthy lifestyles, so reducing the incidence of nutrition-related diseases. Especially important for SARD are improvements in child nutrition to make sure that children grow up able to fulfil useful roles in the society of the future.

6.3 Policies relating to markets and property rights

6.3.1 Resource property rights

The main property rights important for SARD are those relating to land and water (including inland and marine fishery resources).

Typically, most land is held in private or communal ownership, fisheries are in public or communal ownership, or are open access, and most water resources are in public ownership. Policy options in relation to these various types of property rights include:

- policies entailing reallocation of resource property rights between public, communal and private ownership;

- policies relating to the redistribution of privately owned resources among private individuals;

- policies that govern the utilization of open access, state and common property resources; and

- policies to establish and enforce procedures to encourage the efficient and sustainable use of resources.

All four types of policy interventions are likely to have impacts on sustainability, efficiency and equity. Moreover, particularly for the first two, where there is change of ownership, the full impacts of a particular policy may be difficult to predict and may not necessarily be positive for all three criteria. By their very nature, changes in property rights are likely to make some individuals better off and others worse off, meaning, as noted earlier, that such policies are often politically sensitive and divisive.

A change to property rights may be justified when present rights are leading to degradation of an open access or common property resource through over-use. The intervention may range from assistance with the creation or strengthening of an organization of users which has the right to manage the resource, to privatization or nationalization of the property. While the modern tendency is more in favour of the former than the latter, case-by-case consideration of likely implications is required. In Burma and Nepal, forests previously in communal ownership were nationalized with the aim of limiting over-use. The result was an increase in illegal cutting of timber since the dispossessed communities no longer saw the forest as 'theirs', and government enforcement of cutting restrictions was weak. In South Asia, the transfer of irrigation tanks at the time of independence from the control of local rulers to village councils or other democratic bodies generally resulted in a decline in standards of maintenance.

Land settlement schemes are a particular case of the transfer of property rights from public to private ownership, although seldom is the full freehold title to the land given to the settlers. On the other hand, it is increasingly common for governments to sell off property rights, such as the privatization of water supply utilities in Britain.

Very inequitable access to resources may exist when a few individuals own disproportionate

---

Box 8

GOVERNMENT AND VILLAGE PARTNERSHIPS: PROGRAMME NATIONAL DE GESTION DES TERROIRS VILLAGEOIS (PNGTV), BURKINO FASO

Adapted from: Toulmin et al, 1992, quoted by Carey 1994.

Where existing land tenure arrangements are not conducive to SARD, revised legal frameworks can help in the establishment of the necessary conditions for what is called 'primary environmental care'. Primary environmental care fosters natural resource management by building on local skills, local resources and forms of cooperation, and participation to empower local communities. Legal revisions need to be directed to granting rights, access and security of tenure to farmers and pastoralists so as to foster responsibility and farsightedness, and the application of appropriate regulations to prevent pollution and resource degrading activities. A good example is the PNGTV in Burkino Faso.

Under this program, land tenure conditions have been established for widespread action at the local level. It follows the enactment of land tenure reform in 1984, to ensure fair access to land and resources and to encourage greater local involvement in managing and restoring degraded land.

The program is carried out in four stages and now involves about 380 villages, most of which are in the first two stages. First, a village-level committee is established after discussions and training. The committee works with program staff to define and demarcate the village boundaries. A resource inventory is then made. The last two stages involve negotiating and finalizing a contract between the government and the village committee about the investment level for better productivity and management of village resources.

Problems that still need to be resolved include better allocation of formal powers in the village committees and ensuring representation in the committees of all land users, including migrant farmers and herders. In addition, more efficient ways to map, produce resource inventories, and plan land improvements must be found as the current procedure is still too lengthy.

---

shares of a resource. Such situations may be incompatible with SARD and may require a reallocation of ownership. Dispossessed owners may or may not be compensated and, if compensated, may receive full or partial compensation. The most familiar example of policies of this kind is land reform carried out under such slogans as 'land to the tiller'. Experience shows that radical land reform is difficult to implement because of the power and influence of the land-owning class. Moreover, the benefits are often not as great as the proponents hope, at least in the short to medium term. However, these are not reasons for doing nothing when the existing ownership pattern is not sustainable.

Policies to improve the efficiency of use of public or open access property resources include the introduction or variation of prices charged to users, or the imposition of quotas on use. For instance, rights to use common grazing land, forests or fishery resources may be restricted to avoid over-use. In some cases, former curbs on use of common property may have broken down under pressures such as those resulting from population growth and increased commercialization. As a result, policy measures may be needed to strengthen

former rationing institutions or to establish new ones. For example, limits may be set on the number of fishing operators in a given fishery and on the size and nature of each operator's catch.

Water policy issues relate mostly to the private utilization of public, open access or communal resources. Legal and institutional structures governing such utilization vary widely. These structures influence efficiency and sustainability of the use of the resources, including the degradation of the resource by contamination. As urban and industrial use of water grows, the management of water for multiple uses (including re-use) will grow in importance, as will the need for mechanisms to improve the efficiency of water allocation between users. Both realistic water pricing and the introduction, where practicable, of 'polluter pays' charges for water contamination appeal to many economists as 'win-win' policy reforms. They encourage the more sustainable use of a valuable resource and generate financial resources that can be invested in sustainable development, such as watershed management or the maintenance of water catchment infrastructure. However, in both cases, the distributional impacts of the new measures need to be considered carefully. Moreover, the near-impossibility of implementing an effective tax on pollution generated by many small-scale farmers limits the feasibility of applying the polluter pays principle in agriculture.

In a similar fashion, the tendency of many governments to charge loggers less than the full competitive value of the timber in public forests leads to forgone income for the public purse. Worse, it also causes inefficient and unsustainable use of the resource.

Property rights are likely to be a source of inefficiency if they do not provide for the transfer of the resources to those best able to use them. For example, in many countries the outright sale of land is prohibited, and in other cases renting land is illegal. Moreover, the lack of a transferable title to land may inhibit the users of that land from gaining access to formal credit. That will have negative implications for efficiency, and also possibly for sustainability if lack of equity capital precludes investments in land improvements. In such situations, tenure reforms more consistent with SARD might be designed.

Even when land is held in individual tenure, the security of that tenure may affect the care with which the land is conserved and used. If access is for a limited time only, as under some leasehold arrangements, there may be no incentive for the land-user to manage the resource with an eye to long-term productivity. The same may apply when the rules prohibit the inheritance of the property rights by the owner's surviving spouse or children.

Security and transferability of title can be of benefit for other forms of property such as rights of access to common property or state-owned resources like fisheries and irrigation water.

6.3.2 Institutional development

In this context, 'institutions' are defined as the rules, conventions and other elements that form the structural framework of social interaction. They establish the cooperative and competitive relationships that constitute a society and, more specifically, an economic order. Organizations are not the same as institutions under this definition, although an organization may be seen as part of the institutional framework.

There are many different types of institutions potentially important for SARD. They may be divided into community-based institutions and market-based institutions. Within the former are various types of cooperative arrangements and value systems that affect agricultural production and rural resource use. These include informal arrangements such as family ties and responsibilities, systems of rank and social status, shared values and beliefs, religious or other taboos, and peer group pressures. The scope for government policy intervention in these areas is obviously limited, but not negligible. Education and media campaigns can contribute to changed attitudes and values. 'Change agents', such as community leaders, can be 'cultivated' by government officials to pass on information and ideas about sustainable development to others in their communities. Similarly, local organizations such as religious establishments or women's groups may be encouraged by government to promote SARD in local communities.

More formal institutions may be established or manipulated by governments. Examples include laws and regulations governing the operation of cooperatives, political parties and lobby groups, land tenure systems, or rules for the operations of financial organizations such as banks and credit unions. Many of these more formal institutions are mentioned under other headings.

Market-related institutions are concerned with production, processing and marketing of agricultural commodities and with the management of agricultural resources. The category also extends to institutions ancillary to agriculture concerned with the delivery of related services.

Consistent with the notion that command and control systems generally do not perform as well as market-based ones, policies to encourage market development will generally be consistent with SARD. Effective intervention may include strengthening many market-related institutions by removing unnecessary regulations and restrictions on the development of business activity and by the establishment of an economic environment in which business entrepreneurship can prosper. Laws on fair trading may need to be improved, and curbs imposed on the power of monopolies. Provision of good market information systems and other measures to promote the development of small businesses in agricultural marketing will tend to benefit both agricultural producers and food consumers through lower marketing margins. Similarly, the appropriate framework may need to be created to facilitate the provision of finance for a growing rural business sector. Action may also be needed to foster the creation of a stock exchange, if none exists, in order to enable business owners to raise equity capital and spread business risks into the wider community. Despite the usual political sensitivity of such matters, consideration may need to be given to relaxing restrictive rules governing foreign investment in agricultural marketing and finance if local resources are inadequate.

6.4 Policies aimed at establishing democratic and participatory processes

6.4.1 Decentralization

Successful policy making for SARD requires that policy makers have a sound knowledge of the systems of production and resource use with which they are dealing. That usually means a need for a close familiarity with farm and rural household systems—something that may not be easy for urban-based policy makers. Since it is the farmers, foresters, fisherfolk and local people who ultimately have to decide what to do in managing rural resources sustainably, it is logical to involve them as fully as possible in decision making about SARD.

Central policy makers, directors of research and extension and the like cannot have sufficient intimate knowledge of all local circumstances to make top-down policy making work. The more decisions can be decentralized, and the more responsibilities can be delegated to local level, including to local communities, the better the system is likely to work. Of course, it would be irresponsible for important policy decisions to be delegated to local level if, at the same time, local people, organizations and governments were not provided with the training, resources and other support needed to carry out those responsibilities.

Decentralization, by itself, will not work unless clear lines of communication are established between the central authority and the local level. Information about local needs and circumstances must be communicated effectively from the local and provincial level to the centre, so that overall resource allocation decisions can be made sensibly. Also, procedures must be in place to identify where weaknesses at the local or provincial level are impeding progress, so that central authorities can take remedial action.

It follows that the important policy decisions are to determine how far it is practicable and wise to delegate both responsibility and funds to provincial, district and local bodies, including local community groups. Such delegation will require that appropriate ways of monitoring these delegated activities be set up and that effective ways of communicating between the levels are devised and used.

6.4.2 People's participation and empowerment

There is a large literature on people's participation in SARD and it is impossible to do justice to it all here. For the reasons given above, people's participation is an essential element of any successful SARD policy. Unfortunately, however, much of what is written about participation, while long on rhetoric, is short on practical guidelines for implementation.

The basic problem with a policy to promote people's participation is that many existing agricultural organizations, such as universities, research organizations, extension bodies and regulatory authorities, find it difficult to learn from, and work with, farmers and rural people. This is because 'they are characterized by restrictive bureaucracy and centralized hierarchical authority; their professionals are specialists and see only a narrow view of the world; and they have few systematic processes for getting feedback on performance' (Pretty 1995, p. 202). Pretty, among others, has advocated the need for a dramatic paradigm shift based on the definition of new roles for development organizations and professionals, with new concepts, values and behaviours. He does acknowledge, however, that such a revolution will not be easy to achieve.

Given the likely difficulty in achieving the advocated dramatic transformation in procedures for full and effective people's participation, a more pragmatic approach is taken here. A strategy that is evolutionary rather than revolutionary is outlined. It has several components that are discussed in turn.

First, a truly participatory approach requires reasonable development of democratic processes, including respect for human rights, and concern for the status of women, children and minorities. Moreover, participation requires freedom from the ravages of war and

Box 9

CHANGING PROFESSIONALISM FROM THE OLD TO THE NEW

Taken from Pretty 1995, p. 201.67

	From the old professionalism	*To the new professionalism*
Assumptions about reality	Assumption of singular, tangible reality	Assumptions of multiple realities that are socially constructed
Scientific method	Reductionist and positivist; complex world split into independent variables and cause-effect relationships; researchers' perceptions are central	Holistic and post-positivist; local categories and perceptions are central; subject-object and methods-data distinctions are blurred
Strategy and context of inquiry	Investigators know what they want; pre-specified research plan or design. Information is extracted from controlled experiments; context is independent and controlled	Investigators do not know where research will lead; it is an open-ended learning process. Understanding and focus emerge through inter-action; context of inquiry is fundamental
Who sets priorities?	Professionals	Local people and professionals together
Relationship between all actors in the process	Professionals control and motivate clients from a distance; they do not trust people (farmers, rural people etc) who are simply the object of inquiry	Professionals enable and empower in close dialogue; they attempt to build trust through joint analyses and negotiation; understanding arises through this engagement, resulting in inevitable interactions between the investigator and the 'objects' of research
Mode of working	Single disciplinary—working alone	Multi-disciplinary—working in groups
Technology or services	Rejected technology or service assumed to be fault of local people or local conditions	Rejected technology or service is a failure
Career development	Careers are inwards and upwards—as practitioners get better, they are promoted and take on more administration	Careers include outward and downward movement; professionals stay in touch with action at all levels

civil disturbance and reasonable standards of law and order and security.

The extent to which these preconditions are met obviously varies from country to country, and scope for intervention to bring them about when they are lacking may be limited.

Second, there is a need to develop skills and to change attitudes and values both in local communities and among agricultural profes-sionals. SARD issues need to be integrated at all levels of formal and informal education and training. Moreover, according to some

---

Box 10

A TYPOLOGY OF PARTICIPATION IN DEVELOPMENT PROGRAMS AND PROJECTS

Adapted from Pretty 1995, p. 173

1. *Passive participation.* People participate by being told what is going to happen or has already happened. It is a unilateral announcement by administration or project management without any listening to people's responses. The information being shared belongs only to external professionals.

2. *Participation in information giving.* People participate in answering questions posed by investigators using questionnaire surveys or similar approaches. People do not have the opportunity to influence proceedings, as the findings are neither shared nor checked for accuracy.

3. *Participation by consultation.* People participate by being consulted and external agents listen to their views. These external agents define both problems and solutions, but may modify these in the light of people's responses. Such a consultative process does not concede any share in decision making and professionals are under no obligation to accept people's views.

4. *Participation for material incentives.* People participate by providing resources, for example labour, in return for food, cash or other material incentives. Much on-farm research falls into this category, as farmers provide the fields but are not involved in experimentation or the process of learning. Although often called participation, people have no stake in prolonging activities when the incentives end.

5. *Functional participation.* People participate by forming groups to meet predetermined objectives through the development or promotion of externally initiated social organizations. Such involvement is seldom at early stages of project cycles or planning, but rather after major decisions have been made. These institutions tend to be dependent on external initiators and facilitators, but may become self-dependent.

6. *Interactive participation.* People participate in joint analysis, which leads to action plans and the formation of new local institutions or the strengthening of existing ones. These groups take control over local decisions and so people have a stake in maintaining structures or practices.

7. *Self-mobilization.* People participate by taking initiatives to change systems, independent of external institutions. They develop contacts with external organizations for resources and technical advice they need, but retain control over how resources are used. Such self-initiated collective action may or may not challenge existing inequitable distributions of wealth and power.

---

commentators, there is a need to 'transform the learning environment', to change rural organizations into 'learning organizations'. An indication of what these commentators have in mind can be gleaned from Box 9. It is less clear what policy measures are needed to effect such a transformation.

Third, and related to the organizational transformation just mentioned, there is a need to make agricultural research, extension and support systems more sensitive to the needs and circumstances of their clients. A number of ways of conducting research and extension

have been advocated, and some have been tested, based on a participatory approach. While there are often bureaucratic and other impediments to the successful implementation of such methods, a move at least a little way along the road will often be possible. It requires a policy decision in favour of following that route, then that decision to be put into effect, with the necessary reorientation of activities and approaches. Implementation will need to include providing staff with the appropriate training and incentives. In many cases, forging closer links between research and extension will be an important co-requisite for more successful overall engagement with client farmers and rural people.

Fourth, a policy decision to promote and work through local organizations, including NGOs, may circumvent some of the difficulties in transforming hidebound government organizations into people-sensitive entities. This is a matter of delegating both responsibilities and funds to these organizations, and of setting up suitable structures for coordination and consultation, monitoring and evaluation.

Finally, people's participation can be improved by making proper use of appropriate communication and information technology. The two-way flow of information between the rural people and local organizations on the one hand, and policy makers and planners on the other, needs to be as effective as possible. That means using a variety of channels of communication, from local and district meetings and consultations, through to the use of modern electronic media, where appropriate. Finding technological solutions to resource management problems may require local people to be given access, directly or through intermediaries, to relevant electronic or other databases containing information of local relevance. Such information may come from aerial photography or satellite imaging, or may be obtained from the locals themselves. Also relevant is information on prospective improved farming and resource management technologies that have been studied in other, similar environments.

6.5 Policies focussed specifically on natural resource use and environmental protection

While many of the above policies will have significant impacts on rural resource use and the environment, there is a range of policy options that are directly focussed on these areas. These are considered next.

6.5.1 Direct government action

Where, perhaps because of market failure, there is rural resource degradation or contamination of the environment, governments can act directly to try to correct these problems. For instance, governments can fund and undertake land conservation or rehabilitation works. Such works will often be on public land, but may also be on private land, with or without the agreement of the land-holders,

depending on the perceived seriousness of the externalities.

Similarly, governments may opt to bring unused land into use, or under-utilized land into more intensive use, in order to reduce pressure on existing farming areas. This is likely to require works to clear the land, perhaps to shape it, to construct required infrastructure, such as roads, communications systems, irrigation facilities and drainage systems. Many land settlement schemes are based on public investments of these kinds, and often include further public funding to allow the settlers to become established in the new environment.

Both land improvement and land settlement schemes need to be subject to careful prior appraisal, including both environmental impact assessments and social cost-benefit analysis. There are also questions to address about the extent of cost recovery for such works from beneficiaries (or polluters). If the beneficiaries are prosperous, there would seem to be no reason why they should not be required to pay a large share of the cost. However, where relatively poor smallholders benefit, cost recovery may not be appropriate. Moreover, it may not prove possible to devise a cost-effective way of recovering the costs, especially when the beneficiaries are ill-defined and widely dispersed. How are taxes to be raised from the beneficiaries of measures to guarantee cleaner air, for example?

While governments may engage in measures to develop rural production, they also commonly invest in measures to conserve resources, for example by the establishment and management of conservation areas and national parks. Of particular importance for SARD are actions to conserve genetic diversity of useful and potentially useful genetic material. In the case of plants, genetic material may be conserved in such forms as tissue culture, seed banks, plant variety conservation plots, botanic gardens or flora reserves. Animal genetic material is more difficult to conserve. However, small herds or flocks of endangered breeds or species may be kept on research farms or in conservation farms, parks or zoos. Frozen sperm, ova or fertilized embryo may be stored for some time.

Governments may also elect to pay farmers or other agriculturalists to continue to farm endangered types of crop plants or animals, or to maintain flora or fauna reserves on their land.

6.5.2 Control instruments

The range of potential control instruments that governments can consider using is wide. It includes regulations, controls and bans on certain types of resource use or agricultural practices. For example, the use of certain chemicals, such as particular toxic sprays, may be prohibited or strictly controlled. Farmers may be penalized for undertaking certain practices thought to be environmentally damaging, such as burning of crop residues. Alternatively, they may be required, under threat of a penalty, to follow other practices, such as control of noxious pests or weeds.

The allocation of some resources may be controlled by governments or government agencies. For example, it is common for irrigators to be allotted quotas for the amount of water they can draw from a public irrigation system, from a river or from an underground source. Users of a forest resource may be restricted in the number and size of trees they can cut. Increasingly, quotas are being imposed on the size and composition of fish catches, or on the fishing gear that may be used.

As discussed earlier, it is one thing for governments to devise such rules and regulations, but it is not always easy to enforce them. For many individual resource users, the private benefit to be gained by breaking or 'bending' the rules may be considerable. Non-compliance may be on a serious scale. There is also the likelihood that some resource users, especially the more prosperous, will bribe officials to overlook breaches of the regulations. If such corruption becomes widespread, it may even undermine the whole process of orderly government. Moreover, because of the geographically dispersed nature of farming, forestry and, especially, fishing, checking on compliance is likely to be very difficult and expensive. While it is easy for governments to devise rules and regulations for the more sustainable use of rural and fisheries resources, these may not always work as well as the policy makers hope. Therefore, it may be best to look first at ways to encourage sustainability using economic incentives that are not so difficult to implement.

6.5.3 Economic incentives

According to Markandya (1994, p. 19):

> There is a panoply of policy instruments that governments can use to implement an economic incentives approach to environmental management. Some are more applicable to natural resource management, others to environmental protection. Yet others are applicable to both (product and input pricing, taxes, performance bonds, etc). They are based on the same principles of 'getting the prices right' and making property rights clear, secure, and tradeable. Some operate through existing markets (e.g. product pricing), while others imitate the market or seek to create new markets (e.g. tradeable pollution permits). In many cases, such as forest policy and hazardous waste management, it is necessary to employ both sets of instruments to obtain cost-effective and efficient outcomes. An economic incentives package for forest policy would include, at the minimum, concession bidding, resource pricing, resource taxation (royalties), perhaps performance bonds, and investment incentives for replanting.

Markandya then provides a list of the main instruments and a brief description of each. Another, rather more comprehensive list and associated set of descriptions is provided by Panayotou (1994a). For lack of space, these descriptions are presented here in abbreviated form. Some of the listed measures have been mentioned in the discussion above, but others have not. Readers are referred to the original sources for further explanations:

- Tradeable resource shares—the allocation of a transferable right to a percentage share in an 'invisible' or 'uncertain' common property resource, such as a fishery.

- Individual tradeable quotas—the allocation of a transferable quota for a fixed amount of a common property resource.

- Tradeable development rights—the allocation of a right to develop a resource in another area in exchange for relinquishing the right to develop a similar resource in an area of special conservation value.

- Tradeable emission permits—permits that grant polluters the rights over air or water, those rights acquiring an economic value because of the option to sell them to others. An example is the manure permits granted to farmers in the Netherlands.

- Environmental taxes—taxes on environmentally damaging products such as petroleum products and pesticides, designed to provoke a reduction in the use of these products.

- Resource taxes—setting royalties, license fees or other charges on government owned or controlled resources at prices that reflect the true value of the resources.

- Effluent or emission taxes—the familiar 'polluter pays' principle, with charges ideally set in accord with the true cost of the environmental damage caused.

- User charges—setting charges for irrigation water, waste disposal or other services or inputs supplied by the public sector, mainly as a means of cost recovery, but also to reflect something approximating actual costs.

- Deposit refund schemes—presumptive charges on products intended to shift the responsibility for controlling pollution to producers or consumers. Examples are refundable deposits on pesticide containers or vehicle batteries.

- Environmental bonds—economic instruments that ensure adequate funds are available to restore environmental damage, useful, for example, in mining or logging operations.

- Subsidies—financial assistance to promote sustainable behaviour, for example by encouraging investment in resource conservation.

Markandya argues that all the instruments he lists have been used in one country or another, and policy makers concerned with SARD should be familiar with them. Not all are immediately relevant to agricultural and rural resource management, but many are. For instance, allocated water rights can be made tradeable. Moreover, in areas where water supply is uncertain, the tradeable rights might be based on a resource share, as used in some parts of Australia.

On the other hand, there have been difficulties in using some of the suggested measures. In developed countries, setting environmental taxes in such a way as to control pollution effectively and pricing government-owned resources to reflect their true values have both

proved hard to implement. The feasibility of using such measures in less developed countries where administrative systems are less well developed is therefore questionable.

7. INTEGRATED APPROACHES TO SARD

The process of SARD is of such complexity and far-reaching scope that integrated approaches are vital. Actions to promote SARD in one part of the system may fail unless supported by complementary actions in other parts of the system. So, as already noted, efforts to promote SARD within the agricultural sector will not succeed unless the right policies are in place at the macro-level and in other sectors. Integration is also needed across levels of decision making. These levels include the households of producers and consumers, local communities, the business sector, and various levels of government. Similarly, there is a need for integration across sectors, such as agriculture, education, transport and trade, as well as across disciplines and professions. People with different backgrounds and experience must learn to work together, each making his or her contribution while respecting the legitimacy of other viewpoints.

To illustrate these requirements, some features of the needed integrated approaches to SARD in various aspects of agriculture are briefly examined below.

7.1 Improving the efficiency and sustainability of rural resource use

The key components in an integrated approach to improving the efficiency and sustainability of rural resource use include:

- developing human capital through education and training, research and extension and improving information flows;

- land use planning and soil conservation, to prevent, or at least slow, land degradation and to lead to the identification and uptake of sustainable patterns of land use;

- improved water management, through the development and maintenance of infrastructure and the operation of systems of regulating water use to achieve efficient

---

Box 11

POLICY FAILURES FOR SUSTAINABLE FORESTRY

Adapted from Panayotou, 1994b, p. 245.

Forest policy is an excellent example of a resource-specific policy that needs to be overhauled if the link between scarcity and prices is to be established. If indeed we are facing a growing scarcity of forests, forest product prices should be rising to slow down deforestation and accelerate reforestation. At present, not only are most forest products and services not priced, but even timber, which is an internationally traded commodity, is priced below its true scarcity value due to implicit and explicit subsidies and institutional failures. Uncollected resource rents, subsidized logging on marginal and fragile lands, and volume-based taxes on timber removal encourage high grading and destructive logging. Forest concessions are typically too short to provide incentives for conservation and replanting. Failure to value non-timber goods and services results in excessive deforestation, in conflicts with local communities, in loss of economic value and in environmental damage. Promotion of local processing of timber often leads to inefficient plywood mills, excessive capacity, waste of valuable tropical timber and loss of government revenues. Replanting subsidies often end up subsidizing the conversion of valuable natural forests to inferior mono-species plantations, with the associated loss of the value of both tropical hardwoods and biological diversity.

---

allocation between competing uses—ideally through appropriate economic incentives;

- conservation and utilization of biological diversity through appropriate *in situ* and *ex situ* programs;

- development of rural energy supplies through rural energy policies and technologies that promote a mix of fossil and renewable energy sources; and

- improved systems of resource management through better vertical information flows. Such flows may be facilitated by encouraging the fuller participation of rural people in decision making, and by the introduction of better systems for the appraisal, implementation and monitoring of policies, programs and projects.

7.2 Improving the efficiency and sustainability of farm production

The key component in an integrated approach to improving the efficiency and sustainability of farm production is the successful development and adoption of more productive and sustainable technologies. To this end, continued investments in agricultural research are needed, as well as the creation of an environment conducive to innovation by farmers.

The technologies to be introduced must be productive, acceptable to farmers and adapted to their needs and circumstances, and environmentally kind. Some possible approaches include:

- integrated plant nutrition systems that place greater emphasis on biological processes and recycling for the supply of nutrients, so avoiding waste and minimizing nutrient losses that may otherwise pollute water resources;

- integrated pest management systems that are effective in controlling crop and livestock losses while minimizing the use of expensive and potentially hazardous chemicals; and

- integrated grazing systems, particularly for common property grazing lands, to promote efficient forage and livestock production consistent with sustainability.

7.3 Forestry

Measures required for sustainable forestry include:

- Improved forest information systems to allow better policy making and planning.

- Changes in institutional arrangements relating to property rights, the awarding of concessions and the collection of resource

Box 12

THE NEED FOR INTEGRATED SUSTAINABLE DEVELOPMENT OF COASTAL AREAS

Source: FAO, 1994*a*, p. 42.

Two-thirds of marine fish production come from stocks which spend the first and most vulnerable stages of their life cycles in coastal areas. These areas are under serious threat.

Coastal waters are the recipient of eroded soil, pesticides, fertilizers and other pollutants. Coastal area damage is particularly acute in tropical developing countries where both natural and economic conditions contribute to high vulnerability. In these countries, population growth and migration to coastal cities and regions are leading to increases in municipal and industrial discharges and landfill, mangrove clearing, coral mining and other construction-related damage. In some countries, siltation is becoming severe due to deforestation, the construction of lumber roads and land clearing. Intensified agriculture is contributing increasing amounts of pesticides and herbicides to coastal waters.

To tackle the issue, efforts are being made to ensure that plans for the development of coastal areas—both at the area and the sectoral level—integrate all aspects of the problem. This approach is generally known as Integrated Coastal Area Management.

taxes, to prevent over-utilization and to maximize the public benefit from forest utilization.

- Improved management for the multiple use of forests. These uses may include grazing and timber production, fuel-wood and saw-log production, conservation and recreation. Improved management must also recognize and accommodate the customary rights of access and use by local communities.

- Promotion of private forestry through appropriate incentives and financing mechanisms, such as co-financing of long-term loans and good management of long-term risks such as pests, diseases and fire.

- Development and dissemination of technologies for the efficient and sustainable integration of trees into farming systems.

7.4 Fisheries

Responsible fishing means that the activity must be conducted in a way that is ecologically sound and socially just, respecting biological, ecosystem and cultural diversity, in order to guarantee sustainable production.

The attainment of responsible fisheries management will require policies that include:

- the development of a legal framework for fisheries, covering systems of property and use rights, and institutional arrangements to regulate and control the size and nature of catches;

- recognizing the difficulty in monitoring changes in wild aquatic ecosystems, and therefore adopting a precautionary approach to management, for example by setting catch limits at prudent levels;

- the parallel development of measures to educate and encourage local producers and consumers to participate in the management of their marine, lacustrine and riverine ecosystems;

- the introduction of integrated systems of watershed and coastal area management that takes account of the fishery resources in or adjacent to those systems (see Box 12);

- the development of fishing and aquaculture technologies that are productive but consistent with sustainable production and protection of the environment; and

- the negotiation and enforcement of international agreements to protect open access oceanic fish stocks from over-exploitation.

7.5 Rural development

Rural development embraces all the above sorts of measures for the sustainable development of farming, forestry and rural fisheries. However, in addition, SARD will require complementary initiatives in general rural development, including provision of government services for law and order, education, health, etc.

Particularly in countries where agricultural land and water resources are scarce or are of poor quality, there may not be the capacity to develop primary rural production fast enough to provide sustainable livelihoods for all the people. If unacceptable rates of out-migration from these areas are to be prevented, rurally-based industries need to be developed. Rural development policies directed to this end will include:

- Selective and targeted improvements to local infrastructure, including the development of suitable sites for small-scale industries.

- Provision of incentives for investors to locate businesses in rural areas. For example, this may mean making sure that factories or other facilities established in the main urban centres pay the full social costs of locating there, such as the costs of additional congestion and pollution. Providing formal schooling and vocational training for the rural population can also encourage investors to locate businesses in rural areas.

- Provision of appropriate government services in rural areas, including small business development agencies.

8. PROGRAMMING

A 'policy management system' is a strategic and participatory process in which a core management group develops an institutional framework linking national, regional and local initiatives, as well as government departments, private and research sectors, and community-level organizations Carley (1994, p. 19). This system operates within the context of a national sustainable development strategy, as discussed in section 7 above. The strategy is important because policy must be developed and implemented in an uncertain and changing world. Moreover, good policy making is a learning process, able to encompass adaptation in the light of new information, including information from monitoring and evaluating past initiatives, whether successful or not.

Good policy making for SARD will therefore be an iterative, continuous cycle, with the following stages:

- diagnosis of problems and opportunities
- design of possible interventions
- setting the scope of programs or projects
- impact assessment
- appraisal
- decision making
- implementation (action)
- monitoring and evaluation

then back to diagnosis. Not all steps may be followed on every occasion, and there may be many false starts and loops back, as ideas are developed and new information accumulated.

Each main stage is now considered in more detail.

8.1 Diagnosis

8.1.1 Evaluating current situations and policies

Comprehensive guidelines for the conduct of an agricultural sector and policy review (ASPR) are provided in FAO (n.d.). While still provisional, and not written with a specific focus on SARD, those guidelines seem relevant for the diagnostic stage in policy making for SARD. They are too comprehensive to be set out here. However, an ASPR entails a critical review of issues, problems and constraints in the agricultural sector.

Such a review will normally begin with a study of the structure and performance of the agricultural sector. Performance would be judged, so far as possible, by the contributions made by the sector to the aspects of growth, equity, efficiency and sustainability. Indicators of the present position and trends with respect to these aspects will be sought.

Commonly used indicators include measures on contribution to national income. In the context of SARD, it is important to be alert to the well-known limitations of national income accounts, particularly the common tendency to omit or undervalue changes in natural resource stocks and environmental quality. So far as

possible, traditional national income accounts need to be supplemented with environmental and resource accounting information (Ahmad, Serafy and Lutz 1989; Markandya and Perrings 1994).

A framework for the diagnostic phase that may be found useful is the so-called SWOT analysis. A SWOT analysis entails a thoughtful review of the Strengths, Weaknesses, Opportunities and Threats of the agricultural sector for the attainment of SARD.

By the nature of SARD, the diagnostic task needs to be executed taking as long a perspective as is realistic. Problems of underdevelopment or unsustainability of today will, of course, need to be identified, but the policy group also needs to think carefully about future threats to sustainable development that may not yet be evident.

8.2 Design

The design phase will often flow naturally from a successful diagnostic analysis. In terms of the SWOT analysis suggested above, design will build on strengths, mitigate weaknesses, seize opportunities and subvert or overcome threats. For instance, diagnosis may have led to the identification of failures of past policies that need to be corrected. Similarly, where diagnosis has isolated cases of resource degradation or environmental damage due to market failure, it will be natural to look first at opportunities to correct such failure using the policy instruments outlined in sub-sections 6.3.1 and 6.5.3.

Often, but by no means always, policy intervention will require the design of a project or program directed specifically at implementing some initiative to promote SARD. For project design, a range of considerations applies, as set out in detail for irrigation and drainage projects in FAO (1995a). The guidelines provided in that document, and in other sources on project planning, are comprehensive and will not be explored in detail here. Only some aspects of design especially relevant for SARD will be given further attention.

8.2.1 Consultative processes

In the design of policy interventions for SARD, especially in the design of projects and programs, consultation with the resource managers, who will often also be the intended immediate beneficiaries, is important, for all the reasons discussed earlier. Of course, selection of goals and objectives are matters for government. However, when it comes to practical matters of what will and what will not work, mistakes can be avoided by proper consultation with those directly affected (Pretty 1995).

8.2.2 Priority setting

Usually, the diagnostic phase will turn up many more problems and opportunities relating to SARD than it is possible to follow through in terms of designing, funding and implementing appropriate policy interventions within a normal planning period. Some priority setting will therefore be unavoidable.

According to FAO (1991a), special attention should be given to situations where conflicts between demands for environmental protection and agricultural development are most acute. Environmentally endangered agro-ecosystems may often be found to correspond with the loci of most severe rural poverty. However, in setting final priorities it will be necessary to consider:

- the severity and urgency of the actual or threatened problem of poverty and/or resource degradation;

- the feasibility and costs of designing policy measures that will solve, or at least significantly ameliorate, the problem;

- the likely scale of the benefits to be expected from successful intervention; and

- the scope for extending intervention to other areas.

More generally, priority setting for policy interventions needs to based on the extent to which the alternative possible interventions are judged to contribute to the SARD goals of growth, equity, efficiency and sustainability.

Within that broad framework for priority setting, attention needs to be directed to critical areas within the major agro-ecosytems where problems are most likely to arise. The den Bosch Conference outlined a series of priority measures applicable for each of the major agro-ecosystems where critical situations mostly occur. The nominated ecosystems were: drylands and other areas of uncertain rainfall; irrigated lands; humid and per-humid lowlands; mountain and hilly areas; and coastal zones and small islands (FAO 1991a, 22-23 and Appendix 2). A fuller consideration of the types of action required within each of

these major agro-ecosystems was provided in the Conference document: *Strategies for Sustainable Agriculture and Rural Development in Areas with Different Resource Endowments* (FAO 1991f). Readers are referred to this source.

8.3 Setting the scope

An important issue in the design of any project or program is the choice of the scale at which it is to be implemented. There may be a tradeoff to be resolved between opting for a sufficient scale of operation to reap economies of size, and the administrative capacity of the organization responsible for implementation. In addition, there is the possibility that the scale will be constrained by limited funds.

For some policy interventions, nothing short of national-level implementation may be practicable. For example, a substantial increase in the resource taxes that loggers are required to pay would not work if implemented in only one part of the country. Under such arrangements, the logging operators would simply move elsewhere where the taxes were lower. A policy restricting access to a marine fishery may need to be implemented on an international scale to be effective.

At the other end of the scale, local initiatives based on the strong participation of a highly motivated community group may only work if kept to a small scale. In the past, efforts in many parts of the world to expand dramatically the scope of successful local rural development initiatives have often proved disappointing.

8.4 Impact assessment

Almost any policy initiative directed at SARD will have consequences for resource quality and the environment. Some will have significant social impacts. It therefore makes sense to require most such initiatives to undergo an impact assessment before they are adopted. The *natural resource impact* of some policy measure may be defined as consequences that affect the productivity in agricultural use of land, water, and plant and genetic resources. The *environmental impact* refers to consequences that affect the quantity or quality of resources not used in agriculture (Crosson and Anderson 1993, p. 13). The *social impact* of some policy change may be defined as consequences induced in the behaviour and way of life of people affected by the change.

Losses in soil productivity due to erosion are an example of negative natural resource impact. Damage to non-agricultural ecosystems due to uses of agro-chemicals is an example of an environmental impact, and a reduction in the quantity and quality of food consumed is an example of a social impact.

In the case of many policy initiatives, it will not be a trivial task to identify merely the nature of the various impacts of some prospective policy change, ley alone their extent. The difficulty will be greater for general policy measures, such as changes in monetary or fiscal policy, since the impacts of such measures may be quite far-reaching. It is because the implications for sustainability of macro-level policies are so hard to trace that the importance of their impacts is too often overlooked.

Even for policies targeted on SARD, the full implications may be hard to discern. Assessing the magnitudes of the possible impacts of interventions is even harder than merely identifying where the main consequences will arise. Nevertheless, the ability to make good estimates of these impacts is important to the design of sound policies for SARD. A good understanding of the ecology of the farming systems, or of forest of fishery ecology, is obviously vital. While, in most cases, the results of scientific study of the systems will be an important contribution to this knowledge base, the understanding of the people inhabiting the systems will often be at least as important. Certainly, it will be necessary to account for their responses to policy initiatives in impact assessment.

8.5 Appraisal

The main formal tools for the appraisal of policy interventions for SARD are extended cost-benefit analysis, cost-effectiveness analysis and multi-criteria analysis. Each is considered in turn below.

8.5.1 Extended cost-benefit analysis

Conventional cost-benefit analysis (CBA) will be familiar to project economists. However, the basic method has been extended to accommodate concerns about sustainability. Descriptions of extended CBA are to be found in such sources as Dixon and Hufschmidt (1986), Markandya and Pearce (1989), Markandya (1991) and Munasinghe (1993), so that only a brief overview is presented here.

The key concept of extended CBA is the valuation of costs and benefits in terms of *total economic value*, defined as the sum of use value, option value and existence value. *Use value* is the normal economic value assigned to a commodity such as a natural resource. *Option value* is the difference between the discounted expected future use value of an item and the value that the relevant group or society is willing to pay for it today to conserve it for future use. *Existence value* is the measure of value assigned to a commodity by the relevant group to keep it in being, regardless of any possible value the commodity may have in the future.

Valuation methods to assess these various components of total economic value have been developed. Where there is no observable market price, valuation methods must be used. These mostly rely on assessments of 'willingness to pay' for certain features of a commodity or resource, or 'willingness to accept' compensation for the loss of access to those features. These measures may be assessed by reference to related markets, or they may be obtained by various experimental methods.

In the first category, hedonic pricing techniques have been applied to relate prices of a resource or commodity, such as residential land, to observable features of that commodity, such as amenities. Then it is possible to identify the values placed on those amenities in the market, and to extrapolate those value estimates to other situations. An alternative approach is to assign a value to an amenity such as a national park by observing what people actually pay to gain access to that amenity. This usually means measuring their expenditure on travel. Yet a third possibility is to assess the opportunity cost of replacing the resource or amenity with a similar one, or of rehabilitating a damaged or degraded environment. Using this approach, the costs of a forest that is cut down may be valued in terms of the costs to replant and develop another forest that would match the one lost. Similarly, the costs of pollution may be assessed by the costs of clearing up the damage.

Experimental approaches to valuation are based on *contingent valuation* methods. In applying these methods a sample of respondents will be asked hypothetical questions about how much they are willing to pay for a particular commodity, or how much they would need to be paid to accept its loss.

All these methods of valuation have been, and remain, topics for contentious debate amongst economists and others (Carson, Meade and Smith 1993; Desvousges, Gable, Dunford and Hudson 1993; Randall 1993). The same is true of the practice of discounting as normally used in CBA. As discussed earlier, extended CBA is based on discounted costs and benefits, but augmented with a sustainability criterion. This criterion specifies either that the project or program is designed to avoid any overall depletion of the value of the affected resources, or, where this is not possible, that a compensatory investment is made elsewhere to leave the total stock of capital intact.

8.5.2 Cost-effectiveness analysis

It is not always possible to use the valuation methods outlined above. The methods may not be applicable, or may be judged to be too un-reliable, to be used in every situation. In such cases it may be possible to define explicit objectives for the proposed intervention and then to consider the alternative ways in which those objectives might be achieved. Alternatively, a political decision may be made about the objective to be reached, such as stopping damage to the ozone layer, but again alternative ways of attaining that objective exist. In such cases it is possible to compare the alternatives in terms of their costs and determine which is the most cost-effective. Sometimes it may be possible to extend the methods to generate meaningful cost-effectiveness ratios that can be used to aid choice. For example, there may be a number of alternative ways of reducing soil erosion in a particular area that vary in the amount of soil loss that may be prevented. They could be compared in terms of cost per 1000 tonnes of soil loss prevented.

The obvious limitation of cost-effectiveness analysis is that it does not allow comparison of interventions directed at different objectives, so that its applicability in appraising a range of policy measures is very limited.

8.5.3 Multi-criteria analysis

The reality is that most policy interventions for SARD will be directed at the attainment of multiple objectives. Consequently, it will often be difficult, if not impossible, to establish con-vincingly the required valuations to convert all the attributes of some possible intervention to money costs and benefits (van Pelt 1993). Then there is nothing for it but to assess the

alternatives by some form of multi-criteria analysis (MCA) (Pétry 1990).

MCA may be done informally or by using more formal methods. In the latter category there is a range of methods available along with computer software for applying some of them. Typically, MCA involves the following steps:

- Identify the objectives or criteria that are to be allowed to influence the final choice. These should be clearly specified, ideally measurable, and, so far as possible, mutu-ally independent.

- Forecast, for each policy option, the expected levels for each decision criterion. Eliminate any dominated options at this stage. (An option is dominated if it is inferior to some alternative option in terms of all the choice criteria to be considered.)

- Assign a preference measure to each of these criteria levels for each policy option. The preference function may be a propor-tionate score (that is, a linear preference function), or a utility value (that is, a non-linear preference function).

- Assign weights to be applied to the prefer-ence measures for the different criteria. The weighting function may be linear and additive or of some other form. The inter-related nature of the different objectives for SARD may make a linear and additive model misleading, yet the added complex-ity of non-linear models may limit their appeal.

- Calculate the measure of overall value or merit to determine the best option.

While MCA is a flexible method that appears to be well adapted to analysis for policy planning, the complexity and the demands it places on decision makers to be explicit about their objectives and values may limit its use. This is especially so for the theoretically more valid non-linear functional forms. As a result, it may be that only the first three steps above are formalized, followed by intuitive assess-ment of the alternatives.

8.6 Decision making

Choice of the best policy option might seem to follow logically from the previous step of appraisal of alternatives. Yet in practice, the decision maker must confront such difficulties as:

- considerable system complexity;

- imperfect information and hence high uncertainty;

- often, significant downside risk;

- the need to balance several objectives, such as efficiency and resource conservation;

- the need to account for the often conflicting interests of different groups; and

- the need for coordinated action to make solutions work, including actions by others over whom the decision maker has little or no control.

In the face of such difficulties, it is not surprising that some decision makers will seek to 'dodge the issue' by deferring a decision. Unfortunately, inaction is as much a choice as action, and will have its own costs but few benefits. Decision makers must confront the unavoidable complexity of their situation, be prepared to make an informed choice and then be willing to accept responsibility for that decision.

They can often reduce, if not eliminate, the difficulties they face by being clear about objectives, by selecting appropriate criteria to judge the attainment of those objectives, and by getting the best possible forecasts of the values of those criteria under the alternative options. Formal or intuitive assessment of the multiple criteria may then be used to reach a conclusion about what is best, as described in 8.5.3 above.

8.7 Implementation

After a choice, implementation of the decision should follow. Yet difficulties in implementation often arise due to lack of the needed financial or human resources, to imperfect understanding of what needs to be done by those charged with doing it, to failures of inter-agency cooperation, or to many other causes.

Successful implementation of policy decision requires a clarity of purpose, determination by the key decision makers and, above all, adaptability. If impediments cannot be surmounted, maybe they can be circumvented. The notion of SARD as a learning process has already been emphasized, and decision makers must be prepared to make use of what works and to change what does not.

The successful implementation of many SARD policy initiatives requires cooperation among government departments, NGOs and other local organizations, and rural people. Cooperation works best if there has been full participation of all the parties throughout, which is seldom easy to achieve. Cooperation can be encouraged, however, in a number of innovative ways. For example, it is possible to pay funds, not to some central coordinating body, but direct to the parties in return for specified contributions to the cooperative effort. Only small amounts should be allocated at once, with continued funding depending on continued cooperation.

8.8 Monitoring and evaluation

Policies, programs and projects for SARD need to be monitored to see how well they are achieving the objectives set for them and to identify changes needed to enhance progress. Similarly, periodic evaluation of what has been achieved is necessary in order to learn from past successes and failures. Again, the notion of SARD as a learning process emphasizes the need for proper monitoring and evaluation.

Management information systems for monitoring and evaluation need to be developed, consistent with the needs of the case and with the capacity to collect and process information. As in the diagnostic phase, a mix of methods from simple and participatory through 'high tech' may be appropriate.

9. IMPLEMENTING THE GUIDELINES

9.1 Improving information

While there is growing recognition of the importance of SARD and of the significance of the various threats to sustainability, there is still a lack of sound information on these matters. An important priority for the implementation of the procedures set out in this document is to improve the collection, analysis and dissemination of information about SARD, globally, nationally, and locally (Dixon, Hall, Hardaker and Vyas 1994). Information systems such as national income accounts and environmental databases need to be upgraded. There is also a need to produce summaries of the relevant data in timely fashion and in readily understood formats. In these days of computerized data storage and management, and of geographical information systems that include the capacity to generate informative maps and overlays relatively quickly, the scope to answer important questions about SARD is improving. This technology needs to be brought into use in more countries more quickly before impending sustainability problems become serious.

9.2 Education and training needs

The importance of human capital formation for the attainment of SARD has been emphasized at several points in these guidelines. To a large extent, that means there is a need to step up relevant programs of education and training (Boddington 1996).

SARD is everyone's concern, so, in principle, everyone needs to be educated about it and trained to do their part in improving the situation. In reality, some priorities need to be set for the provision of relevant education and training. High priority target groups include general and rural policy makers and analysts; agricultural research and extension personnel; and rural opinion leaders.

In the longer term, the reality needs to be recognized that it is rural people themselves who are responsible for most rural resource management. Therefore, they are the ones, above all, whose skills need to be improved. That may be approached in a variety of ways—gradually, though the education of children in the schools, through adult education and training programs, and indirectly, through rural elites, such as community leaders and government officials.

9.3 International cooperation

In the last analysis, sustainability is a global issue. Therefore, international cooperation is vital. Trade reform should encourage the more sustainable use of resources by leading to the location of production according to the principle of comparative advantage. Foreign aid and improved international financing can give governments and individuals access to the funds needed for investment in the future of agriculture and rural industries. International cooperation is needed in important areas such as the management of oceanic fish stocks, the limitation on greenhouse gas emissions, or to control the international spread of pests and diseases. International agencies such as FAO can contribute to the increased awareness of governments about both their domestic and

international obligations in the pursuit of SARD.

9.4. Future Directions for Research on SARD

Before SARD, there was simply ARD, with its features of efficiency, equity, and poverty reduction. SARD, as the name suggests, grafts on the additional concept of sustainability. It is generally taken that efficiency, equity, and poverty reduction are important conditions of sustainability. However, there may exist very complex trade-offs between sustainability and the other criteria of ARD. Analysis of the nature of these tradeoffs from a policy making standpoint would make for an interesting future research agenda.

If these tradeoffs are significant, it would then be clear that there are gainers and losers from SARD compared to ARD. If many real world "policies" are not sustainable, it is not simply a result of policymakers lacking proper guidelines for producing sustainable policies, but that there is a political economy aspect to the implementation of SARD policies that needs to be considered. However, that little work has been done on to date on this aspect. Future work can start by mapping the conditions under which, for instance, feasible coalitions in support of SARD policies could be put into place, and identify as well which approaches to SARD have greater likelihood of meeting the political feasibility criterion. All of this is difficult to do, and largely left to be done.

10. REFERENCES

Ahmad, Y.J, Serafy, S.E and Lutz, E. (eds) 1989, *Environmental Accounting for Sustainable Development*, The World Bank, Washington, D.C.

Baumol, W.J and Oates, W.E. 1988, *The Theory of Environmental Policy*, second edn, Prentice-Hall, Englewood Cliffs, N.J.

Beckerman, W. 1992, 'Economic growth and the environment: whose growth? Whose environment?' *World Development* 20(4), 481-96.

Boddington, M. 1996, 'Training issues for sustainable agriculture and rural development', in S.A. Breth (ed.), *Integration of Sustainable Agriculture and Rural*

Development Issues in Agricultural Policy, Winrock International Institute for Agricultural Development, Morrilton, Arkansas.

Bromley, D.W. and Cernea, M.M. 1989, *The Management of Common Property Resources*, Discussion Papers 57, The World Bank, Washington, D.C.

Carley, M. 1994, *Policy Management Systems for Sustainable Agriculture and Rural Development*, International Institute for Environment and Development and Food and FAO, Rome.

Carson, R.T., Meade, N.F. and Smith, V.K. 1993, 'Contingent valuation and passive-use values: introducing the issues', *Choices*, Second Quarter, 5-8.

Chambers, R. and Conway, G.R. 1992, *Sustainable Rural Livelihoods: Practical Concepts for the 21st Century*, Discussion Paper 296, Institute of Development Studies, Falmer, Sussex.

Chissano, J.A. 1994, 'Natural resource management and the environment: widening the agricultural research agenda', *Annual Report 1993*, ISNAR, The Hague.

Crosson, P. and Anderson, J.R. 1993, *Concerns for Sustainability: Integration of Natural Resource and Environmental Issues for the Research Agendas of NARS*, Research Report 4, ISNAR, The Hague.

Crosson, P. and Anderson, J.R. 1995a, *Achieving a Sustainable Agricultural System in Sub-Saharan Africa*, Building Blocks for Africa 2025, Paper No. 2, AFTES, The World Bank, Washington, D.C.

Crosson, P. and Anderson, J.R. 1995b, Degradation of resources as a threat to global agriculture, unpublished paper, Resources for the Future and The World Bank.

de Haen, H. and Saigal, R. 1992, 'Agricultural growth and the environment: trade-offs and complementarities', *Quarterly Journal of International Agriculture* 31(4), 321-39.

Desvousges, W.H., Gable, A.R., Dunford, R.W. and Hudson, S.P. 1993, 'Contingent valuation: the wrong tool to measure passive-use losses', *Choices*, Second Quarter, 9-11.

Dixon, J. A., and Hufschmidt, M. M. 1986, *Economic Valuation Techniques for the Environment: A Case Study Workbook*, Johns Hopkins. Baltimore.

Dixon, J.M., Hall, M., Hardaker, J.B. and Vyas, V.S. 1994, *Farm and Community Information Use for Agricultural Programmes and Policies*, FAO Farm Systems Management Series No. 8, Food and Agriculture Organization of the United Nations, Rome.

Edwards-Jones, G. 1996, *Environmental Impact Assessment*, Proceedings of the Training for Trainers Seminar on Environmental and Sustainability Issues in Agricultural Policy Analysis and Planning, Cyprus, 8-19 May 1995, FAO, Rome.

FAO. 1989, 'Sustainable development and natural resources management', Twenty-Fifth Conference, Paper C 89/2 - Sup. 2, Food and Agriculture Organization, Rome.

FAO. 1991, FAO/Netherlands Conference on Agriculture and the Environment, 'S-Hertogenbosch, The Netherlands, 15-19 April, 1991, FAO, Rome.

a   *Elements for Strategies and Agenda for Action (Draft Proposal).*

b   *Resumés des Documents de Base.*

c   *The den Bosch Declaration and Agenda for Action on Sustainable Agriculture and Rural Development: Report of the Conference.*

Main Documents:

d   1. *Issues and Perspectives in Sustainable Agriculture and Rural Development.*

e   2. *Technological Options and Requirements for Sustainable Agriculture and Rural Development.*

f   3. *Strategies for Sustainable Agriculture and Rural Development in Areas with Different Resource Endowments.*

g   4. *Criteria, Instruments and Tools for Sustainable Agricultural and Rural Development.*

Background Documents

h 1. *Sustainable Development and Management of Land and Water Resources.*

i 2. *Sustainable Crop Production and Protection.*

j 3. *Livestock Production and Health for Sustainable Agriculture and Rural Development.*

k 4. *Farming, Processing and Marketing Systems for Sustainable Agriculture and Rural Development.*

l 5. *Social and Institutional Aspects of Sustainable Agriculture and Rural Development.*

Regional Documents:

m 1. *Sustainable Agriculture and Rural Development in Sub-Saharan Africa.*

n 2. *Sustainable Agriculture and Rural Development in Asia and the Pacific.*

o 3. *Sustainable Agriculture and Rural Development in Latin America and the Caribbean.*

p 4. *Sustainable Agriculture and Rural Development in the Near East.*

Miscellaneous Documents

q 1. *Nutrition and Sustainable Agriculture and Rural Development.*

r 2. *Population, Environment and Sustainable Agriculture and Rural Development.*

s 3. *Legal Aspects of Sustainable Agriculture and Rural Development.*

t 4. *Climate Change and Agriculture, Forestry and Fisheries.*

u 5. *Forestry and Food Security.*

v 6. *A New Approach to Energy Planning for Sustainable Agriculture and Rural Development.*

FAO. 1993, *The Challenge of Sustainable Forest Management*, FAO, Rome.

FAO. 1994*a*, *Strategies for Sustainable Agriculture and Rural Development: New Directions for Agriculture, Forestry and Fisheries*, FAO, Rome.

FAO. 1994*b*, *The Road from Rio: Moving Forward in Forestry*, FAO, Rome

FAO. 1995*a*, *Guidelines for Design of Irrigation and Drainage Investment Projects*, (External Draft Review), FAO, Rome.

FAO. 1995*b*, *Responsible Fisheries*, DEEP (Development Education Exchange Papers), FAO, Rome.

FAO. 1995c, *Forest Resources Assessment 1990*, Forestry Paper 124, FAO, Rome.

FAO. n.d., *Outline for Guidelines for FAO Policy Assistance in the Agricultural Sector*, FAO, Rome.

Fisher, A..C. and Hanemann, W. 1990, 'Information and the dynamics of environmental protection: the concept of the critical period', *Scandinavian Journal of Economics* 92(3), 399-414.

IFPRI. 1995, *A 2020 Vision for Food Agriculture and the Environment: The Vision, Challenge, and Recommended Action*, International Food Policy Research Institute, Washington, D.C.

Lipton, M. 1989, 'New strategies and successful examples for sustainable development in the Third World', *Testimony presented at a hearing on Sustainable Development and Economic Growth in the Third World held by the Joint Economic Committee of the U.S. Congress, Subcommittee on Technology and National Security*, June 20, 1989.

Little, I.M.D. and Mirrlees, J.A. 1974, *Project Appraisal and Planning for Less Developed Countries*, Heinemann, London.

Ludwig, D., Hilborn, R. and Walters, C. 1993, 'Uncertainty, resource exploitation, and conservation: lessons from history', *Science* 260, 17, 36.

Lynam, J.K. and Herdt, R.W. 1989, 'Sense and sustainability: sustainability as an objective in international agricultural research', *Agricultural Economics* 3, 381-98.

McCalla, A. 1994, *Agricultural and Food Needs to 2025: Why We Should Be Concerned*, Consultative Group on International Agricultural Research, Washington, D.C.

Markandya, A. 1991, *The Economic Appraisal of Projects: The Environmental Dimension*, Inter-American Development Bank, Washington, D.C.

Markandya, A. 1994, 'Criteria, instruments and tools for sustainable agricultural development', in A. Markandya (ed.), *Policies for Sustainable Development: Four Essays*, Economic and Social Development Paper 121, FAO, Rome, 1-70.

Markandya, A. 1995, *The Integration of Environmental and Sustainability Considerations in Agricultural and Rural Development Policy and Planning: Review and Proposals for Training*, Internal Document 25, TCAS, FAO, Rome.

Markandya, A., and Pearce, D. W. 1989, *Environmental Policy Benefits: Monetary Valuation*, OECD, Paris.

Markandya, A., and Pearce, D.W. 1991, 'Development: the environment and the social rate of discount', *World Bank Research Observer* 6(2), 137-52.

Markandya, A., and Perrings, C. 1994, 'Resource accounting for sustainable development: Basic concepts, recent debate, and future needs', in A. Markandya (ed.), *Policies for Sustainable Development: Four Essays*, FAO Economic and Social Development Paper 121, FAO, Rome, 71-151.

Markandya, A., and Richardson, J. 1994, 'Macroeconomic adjustment and the environment', in A. Markandya (ed.), *Policies for Sustainable Development: Four Essays*, FAO Economic and Social Development Paper 121, FAO, Rome, 153-204.

Meadows, D.H. et al. 1972, *The Limits to Growth: A Report of the Club of Rome's Project on the Predicament of Mankind*, Universe Books, New York.

Munasinghe, M. 1993, *Environmental Economics and Sustainable Development*, World Bank Environment Paper No. 3, The World Bank, Washington, D.C.

Munasinghe, M. and Cruz, W. 1995, *Economywide Policies and the Environment: Lessons from Experience*, World Bank Environment Paper No. 10, The World Bank, Washington, D.C.

Panayotou, T. 1994a, *Economic Instruments for Environmental Management and Sustainable Development*, Environmental Economics Series Paper No. 16, Environment and Economics Unit, United Nations Environment Programme, Nairobi.

Panayotou, T. 1994b, 'Economic instruments for natural resources management in less developed countries', in A. Markandya (ed.), *Policies for Sustainable Development: Four Essays*, FAO Economic and Social Development Paper 121, FAO, Rome, 205-68.

Pétry, F. 1990, *Multicriteria Decision Making and Rural Development*, TCAS, ID No. 13, FAO, Rome.

Pétry, F. 1995a, *Sustainability Issues in Agricultural and Rural Development, Vol 1, Trainee's Reader*, Training Materials for Agricultural Planning 38/1, FAO, Rome.

Pétry, F. 1995b, *Sustainability Issues in Agricultural and Rural Development, Vol 2, Trainer's Kit*, Training Materials for Agricultural Planning 38/2, FAO, Rome.

Pezzey, J. 1992, *Sustainable Development Concepts: An Economic Analysis*, World Bank Environment Paper No. 2, The World Bank, Washington, D.C.

Pretty, J. 1994, 'Alternative systems of inquiry for a sustainable agriculture', *IDS Bulletin* 25(2), 37-49.

Pretty, J.N. 1995, *Regenerating Agriculture*, Earthscan, London.

Randall, A. 1993, 'Passive-use values and contingent valuation—valid for damage assessment' *Choices*, Second Quarter, 12-15.

Schuh, G.E. and Archibald, S. 1996, 'A framework for the integration of environmental and sustainable development issues into agricultural planning and policy analysis in less developed countries', in S.A. Breth (ed.), *Integration of Sustainable Agriculture and Rural Development Issues in Agricultural Policy*, Winrock International Institute for Agricultural Development, Morrilton, Arkansas.

van Pelt, M.J.F. 1993, *Sustainability-Oriented Project Appraisal for Less Developed Countries*, Thesis Wageningen.

von Braun, J. 1992, 'Agricultural growth, environmental degradation, poverty and nutrition: links and policy implications',

Quarterly Journal of International Agriculture 31(4), 340-63.

WCED. 1987, *Our Common Future: The Bruntland Report*, Oxford University Press from the World Commission on Environment and Development, New York.

World Bank. 1991, *Environmental Assessment Sourcebook.* 3 vols, The World Bank, Washington, D.C.

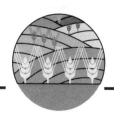

1. Searching for common ground – European Union enlargement and agricultural policy, 1997

2. Agricultural and rural development policy in Latin America – New directions and new challenges, 1997

3. Food security strategies – The Asian experience, 1997

4. Guidelines for the integration of sustainable agriculture and rural development into agricultural policies, 1997